T0331885

Wireless Public Safety Networks 3

Series Editor
Pierre-Noël Favennec

Wireless Public Safety Networks 3

Applications and Uses

Edited by

Daniel Câmara
Navid Nikaein

First published 2017 in Great Britain and the United States by ISTE Press Ltd and Elsevier Ltd

ISTE Press Ltd
27-37 St George's Road
London SW19 4EU
UK

www.iste.co.uk

Elsevier Ltd
The Boulevard, Langford Lane
Kidlington, Oxford, OX5 1GB
UK

www.elsevier.com

British Library Cataloguing-in-Publication Data
A CIP record for this book is available from the British Library
Library of Congress Cataloging in Publication Data
A catalog record for this book is available from the Library of Congress
ISBN 978-1-78548-053-9

Printed and bound in the UK and US

Contents

Preface

Book overview

This third book on Public Safety Networks (PSNs) aims to explore future developments and tendencies in the disaster management field, and to evidence some real-use cases of applications.

Chapter 1 presents an overview of public warning systems and receivers, and their applications. The emphasis is on the victims' point of view and the requirements linked to public warning systems. It also presents the results of two European projects dedicated to the early warning of citizens, namely Alert4All and PHAROS.

Chapter written by Daniel CÂMARA and Navid NIKAEIN.

Chapter 2 introduces a method to develop cost-effective and collaborative early warning systems where sensing and communication capabilities of end-user smartphones can be used to detect and automatically spread information about seismic events. The proposed architecture provides the means to automatically send alerts immediately after an event to all stakeholders.

Chapter 3 presents a real use-case of the SmartSafe project, which is being developed in Ecuador, with the objective of minimizing the rescue time of kidnapping. The proposed distributed system can, in real time, notify the nearby community about an instance of kidnapping.

Chapter 4 describes new strategies for collecting data with the objective of creating real-time localized crisis maps. This information is fundamental to accessing the present situation on disaster sites and organize rescue operations.

Chapter 5 focuses on the development of applications that are context-centric. In an emergency, it is fundamental that the shared information takes into consideration the present situation and the current context. The proposed context-aware middleware is capable of aggregating, compiling and communicating contextual information to improve people's capability to handle emergency situations.

Chapter 6 provides an overview of the different scenarios that can be identified in emergency situations such as natural disasters or accidents as well as the new services that might be beneficial for these scenarios. It also discusses the major challenges to be faced by PSNs to maintain communications in emergency situations. Additionally, a new video hardware platform to be used by first responders is introduced alongside new components to be included in the operators' networks to provide new services in the area of Public Safety (PS).

Chapter 7 presents an overview of one of the most promising technologies in the PS field, that is, the use of flying equipment in rescue operations. It not only presents how these platforms can be used to provide communication to disaster sites, but also provides the details of the air-to-ground propagation model.

Chapter 8 discusses the problem of maintaining a consistent topology for autonomous drones in the context of search and rescue operations. The objective of a self-organizing topology control mechanism for autonomous drones is to provide a steady and consistent architecture to enable the development of upper-layer protocols and applications.

Chapter 9 focuses on a method to formally verify the development of protocols and methods in the context of PSNs. The use of autonomous systems in PSNs is a mark of modern emergency management; however, it is really important to ensure both safe and secure execution of deployed systems.

Chapter 10 centers its discussion on the practical use of *ad hoc* communication in disaster sites. Opportunistically formed networks are far more resilient to disasters since they are physically detached from any infrastructure components, but the real deployment of these networks in disaster situations is an extremely challenging task.

Chapter 11 reviews the potential of Delay Tolerant Networks (DTN), Information Centric Networks (ICN) and Mobile Edge Computing (MEC) to provide valuable services for PS applications. It also discusses how a DTN/ICN PS application can be automatically deployed at the network edge by a LTE base station operating in the disconnected core scenario. An architecture is proposed to transform a disconnected base station into a PS infrastructure, providing communication services in a disastrous situation.

1

Public Warning Applications: Requirements and Examples

1.1. Introduction

Efficient emergency management requires authorities to be able to timely communicate with citizens at risk and provide them with the relevant information about risks and on-going emergencies as well as with the recommended protective actions. The current development of communication technologies together with the existence of newly developed personal receiver devices, such as smartphones, tablet PCs and navigators, makes it possible to provide authorities (alert message issuers) with end-to-end communication means towards citizens at risk (alert message recipients). Moreover, the processing power and storage capacity available in these receiver devices allow multi-channel public warning systems to make use of narrowband channels, such as the ones available in satellite navigation systems, moving the complexity of the system to the receiver devices.

Chapter written by Javier MULERO CHAVES and Tomaso DE COLA.

This chapter presents an overview of the communication processes between authorities and citizens at risk which are present in emergency management. Thereafter, the chapter presents an overview of public warning systems from the recipient's point of view, focusing on different communication technologies and receiver devices which can be used. Finally, the chapter identifies the requirements that public warning applications shall fulfill, based on the results of the European Alert4All and PHAROS projects and provides a description of the public warning applications developed in the context of these projects.

1.2. Emergency management communications

When discussing emergency management in general and, in particular, emergency management communications, two main categories of actors must be taken into consideration: authorities and citizens. The former group is formed by the different (generally public) entities in charge of managing emergencies within a given area of responsibility, including but not limited to fire brigades, civil protection, medical emergency services and police departments. The emergency management responsibilities held by the different entities differ from country to country and even from region to region within the same country; therefore, the entire group of authorities will be treated as a whole within this chapter. The latter are the citizens who, on the one hand, might be at risk of being affected by the emergency and, on the other hand, might actively participate in the management of the emergency situation, for instance, by providing information to the authorities.

During the emergency management process, several types of interaction can take place between authorities and citizens in different phases of the emergency management cycle, as shown in Figure 1.1. For each of the cases, different communication tools and systems are used by the actors, and different communication strategies are put in place, as depicted in Table 1.1.

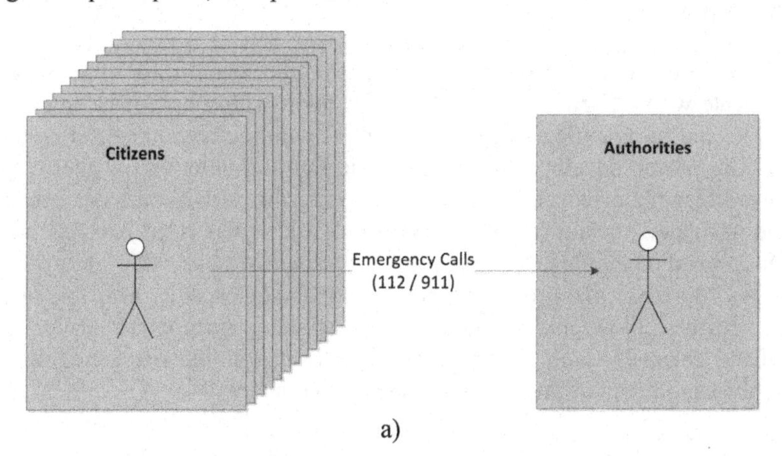

a)

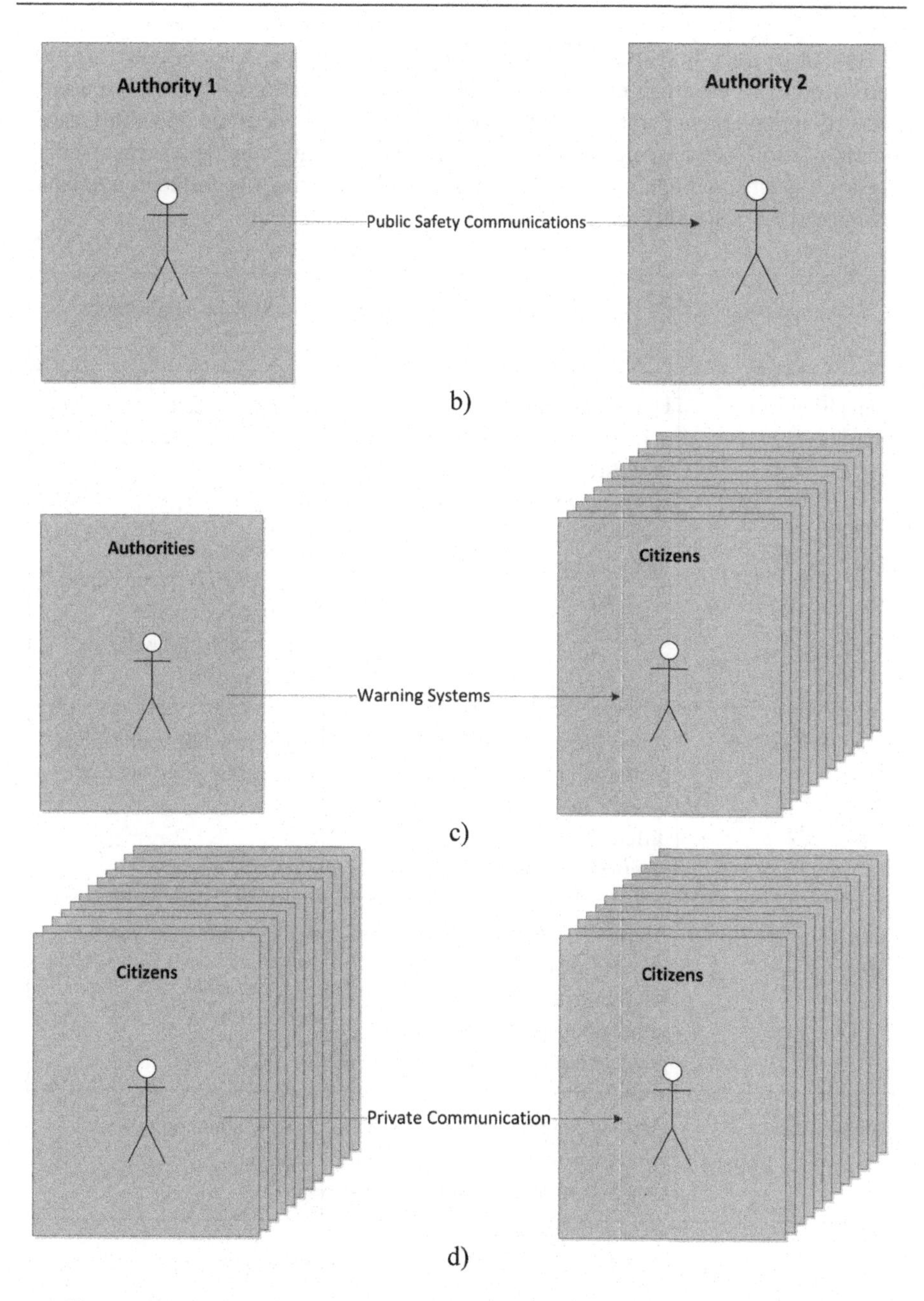

Figure 1.1. *a) Communication from citizens to authorities; b) communication among authorities; c) communication from authorities to citizens and d) communication among citizens*

As can be seen in Table 1.1, a wide range of systems and applications can be used in order to establish efficient and robust communications between the actors involved in emergency management. This chapter will focus on the third case, communication between authorities and citizens, detailing the requirements that warning systems and applications shall satisfy and analyzing the different available options and the suitability of the solutions.

Communication flow	Characteristics	Systems/applications
From citizens to authorities	Citizens generally address authorities in order to notify them of the detection of a possible risk or emergency to provide information about an ongoing emergency or thread or to ask for help	Emergency calls (e.g. 112 in Europe or 911 in North America) Social Networks (e.g. contacting the official profiles of civil protection authorities or police departments in Twitter or Facebook)
Among authorities	Authorities use a wide range of secure and robust communication systems in order to coordinate the different teams and exchange emergency-related information	Voice and data applications over public safety networks
From authorities to citizens	Authorities address citizens at risk in order to provide them with relevant emergency-related information and the recommended protective actions to be put in place	Public warning systems
Among citizens	Citizens communicate among them in order to share information about expected or on-going emergencies	Private communication systems

Table 1.1. *Emergency communication flow*

1.3. Public warning systems

An Early Warning System (EWS) represents the set of capacities needed to generate and disseminate timely and meaningful warning information that enables at-risk individuals, communities and organizations to prepare and act appropriately and in sufficient time to reduce harm or loss [UNI 09]. Based on this definition, the International Federation of Red Cross and Red Crescent Societies have identified the four core components of an EWS [INT 12], namely:

– risk knowledge to build the baseline understanding about the risk;

– monitoring to identify how risks evolve through time;

– response capability;

– warning communication which packages the monitoring information into actionable messages understood by those that need, and are prepared, to hear them.

This includes, on the one hand, the gathering, processing and presentation of information in a consistent and meaningful manner to allow the generation of alert messages and, on the other hand, the generation and transmission of alert messages to the citizens at risk by means of warning communication. Therefore, from an operational perspective, Public Warning Systems (PWSs) can be divided into two main functional modules: an information aggregator, which provides risk knowledge and monitoring functionalities, and an alert dispatcher, which makes use of the available response capability and warning communication to reach the citizens at risk [PÁR 16]. Taking the interaction between Public Warning Systems (PWSs) and public warning actors defined in [PÁR 16] as the starting point, a simplified version, focusing on the communication between PWS and alert message recipients (citizens at risk), can be seen in Figure 1.2. This chapter will focus on the alert (message) dispatcher functionalities and the applications which can be used at the recipient side in order to efficiently receive, decode (if needed) and present alert messages to alert recipients and allow them to understand the situation and put into practice, if required, the recommended protective actions.

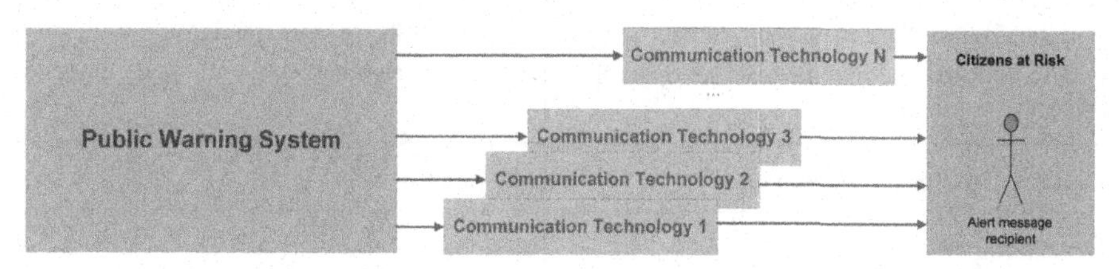

Figure 1.2. *Interaction between PWS and alert message recipients*

As it can be seen in Figure 1.2, there is a wide range of communication technologies that can be used to disseminate alert messages towards the population at risk. Each communication technology used for that purpose provides a different set of features with respect to others and, at the same time, influences the efficiency of the dissemination of alerts [MUL 14]. Additionally, different communication technologies require alert recipients to have or to access dedicated receiver devices for the reception of alert messages over the available bearer services. A summary of communication technologies and the related receiver devices, if needed, is described in Table 1.2.

Communication technology/bearer service	Receiver device
Sirens	Not needed
Tannoy	Not needed
Information billboards	Not needed
SMS	Cell phone/smartphone
Cell broadcast	Cell phone/smartphone
Internet access (either provided through terrestrial, mobile wireless or satellite data networks) using dedicated applications	Cell phone/smartphone Portable and fixed devices (desktop computer, tablet PCs, laptops, etc.)
Pager	Pager
TV broadcast (either satellite, terrestrial or cable TV)	Television receiver
Hybrid Broadband Broadcast TV (HbbTV)	HbbTV-enabled receiver
Radio Data Service (RDS)	RDS-enabled radio receiver
Satellite-Based Augmentation Systems (SBAS)/Global Navigation Satellite System (GNSS)	SBAS/GNSS receiver (navigation device/smartphone
Building notification system	Not needed
Evacuation systems	Not needed

Table 1.2. *Communication technologies and receiver devices*

The efficiency of the transmitted alert message is affected by both the communication technology used and the corresponding receiver device. With regard to receiver devices, a first classification can be done between communication technologies which require citizens to have a dedicated receiver device and the ones which do not. Firstly, communication technologies which allow the reception of alert messages without the need for a dedicated receiver device can have a higher penetration, since the alert message recipient does not need to take any action in order to receive the message, but on the other hand, the penetration of alert messages depends strongly on the location of alert distributors (sirens, evacuation systems, building notification systems, tannoys, electronic billboards, etc.). Therefore, this type of solution is very effective and widely used in specific risk locations, such as highways, chemical plants or power plants, but is not that efficient for the distribution of alert messages in wider areas. Secondly, in cases where a specific receiver device is needed, portability of the device and current behavioral trends play an important role in the effectiveness of alert message distribution. In this regard, two main categories of devices can be identified: portable and non-portable devices [MUL 14].

The first category of receiver devices includes a wide range of personal communication devices (smartphones, cell phones, tablet PCs, pagers, etc.) which allow the reception of alert messages through a wide range of communication technologies, such as wireless mobile networks, terrestrial networks and navigation satellites (SBAS/GNSS). The computational power provided by current devices, as well as the available storage capacity they can provide, allows the transmission of efficiently encoded alert messages and the presentation of alert messages to citizens at risk in different languages and modes, for instance text or voice, thus addressing citizens with special cognitive needs. The use of this type of devices allows end-to-end transmission of alert messages from the alert message issuer to the alert message recipient.

The second category includes devices which have traditionally been available in households to be used by the entire community, like television and radio receivers (although portable TVs and radio receivers are available in the market). The effectiveness of alert messages received using these devices is affected by technical and behavioral aspects. On the one hand, alert messages distributed through TV or radio are generally received by the TV or radio broadcaster, which has to further process the message and transmit it to the alert message recipient (generally embedded in the TV or radio news or using subtitles in the TV case). This limitation has been overcome in recent years by the use of so-called “smart TVs”, which could allow commercial TVs to receive alert messages via an Internet connection. Similarly, the use of the Hybrid Broadband Broadcast TV (HbbTV) would allow the reception of alert messages either in the data carousel of the TV signal or using the Internet connection provided. Decoding and presentation of the received alert messages could be done in both cases, thanks to dedicated applications available in receiver devices [PFE 13]. On the other hand, from a non-technical perspective, penetration of alert messages using this category of devices is limited by the fact that alert message recipients must be using receiver devices at the moment when the message is received.

1.4. Public warning applications

Taking into account the wide range of communication technologies and receiver devices used for the dissemination of alert messages, dedicated public warning applications can be used for receiving, decoding and presenting alert messages to the alert message recipient. Warning applications, in a general case, can be applied for end-to-end transmission between the alert message issuer and the alert message recipient, as shown in Figure 1.3 (adapted from [PÁR 16]), regardless of the communication technology being used.

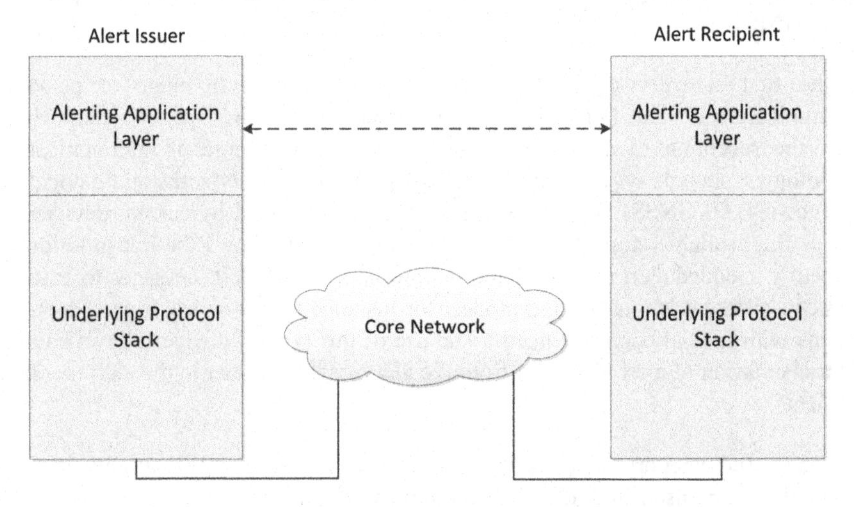

Figure 1.3. *Communication between the alert issuer and the alert recipient*

While the system considerations at the issuer side have been discussed and detailed in [PÁR 16], this section will discuss the main features to be provided by public warning applications at the recipient side, identifying the requirements to be fulfilled. Table 1.3 shows a list of user requirements to be taken into account for the design and implementation of public warning applications. These requirements are an adaptation of the ones identified in the framework of the European Alert4All [ALE 13] and PHAROS [PHA 16] projects.

Req. no.	Requirement description
1	Citizens at risk shall be able to receive alert messages in a timely manner
2	Citizens at risk shall be able to receive consistent alert messages through different alerting channels (communication technologies)
3	Citizens at risk shall be able to receive alert messages during the whole emergency management cycle
4	Citizens at risk shall be able to receive alert messages using standard commercial receiver devices
5	Citizens at risk shall receive relevant alert messages according to their current location (geo-fencing)
6	Citizens at risk shall be able to clearly understand the received alert messages and identify the relevant information about the risk: hazard type, severity, urgency, likelihood, timing and recommended protective action
7	Citizens at risk shall be able to receive alert messages which clearly demonstrate that they have been issued by a recognizable authorized source
8	Citizens at risk shall have the ability to recognize when an alert message is a new one, an update related to any previous one or a repetition of a previous alert message
9	Citizens with any cognitive special need shall be carefully considered to ensure that they receive and understand alert messages
10	Citizens at risk shall be able to receive alert messages in a language they can understand
11	Citizens at risk shall be able to get the alert message presented in different modes apart from text, such as speech or videos, according to their configuration
12	Citizens at risk shall be able to configure several features of their receiver applications, such as the language and mode in which alert messages shall be presented
13	Citizens at risk shall be able to identify in the alert message where to find additional information about the risk

Table 1.3. *List of user requirements related to public warning applications*

Taking the presented requirements into consideration, the main features to be provided by public warning applications can be grouped into the following three main areas:

– Decoding: in cases using any type of encoding to transmit the messages over a particular communication technology, the corresponding decoding shall be applied at the recipient side. Different types of encoding can be applied for an efficient transmission of messages in multi-channel approaches. The purpose of applying encoding techniques is generally to improve the security of the communication between the issuer and the recipient and/or to allow the transmission of alert messages over narrowband channels, reducing the capacity required and the transmission delay (e.g. in the case of SBAS/GNSS channels) [DEC 12a];

– Composition: as a previous step before their presentation to the recipient, a human-readable version of the alert message must be composed, following the composition rules (syntax) of the corresponding presentation language;

– Presentation: according to the recipient's preferences, alert messages can be presented in different languages and modes. Public warning applications shall allow recipients to configure the language in which the message should be presented and the corresponding mode (text, speech, icons, videos, etc.). Multi-modal presentation of alert messages can improve alert message efficiency, by increasing the understanding of alert messages for people with special cognitive needs [JOH 10, LAN 10, SUL 10].

Based on these three main functions, Figure 1.4 shows a high-level operational architecture for public warning applications at the recipient side. In the figure, the three main building blocks are depicted, together with the database containing the citizen's configuration, which shall be taken into account in order to present the alert messages in the language and mode selected by the corresponding citizen.

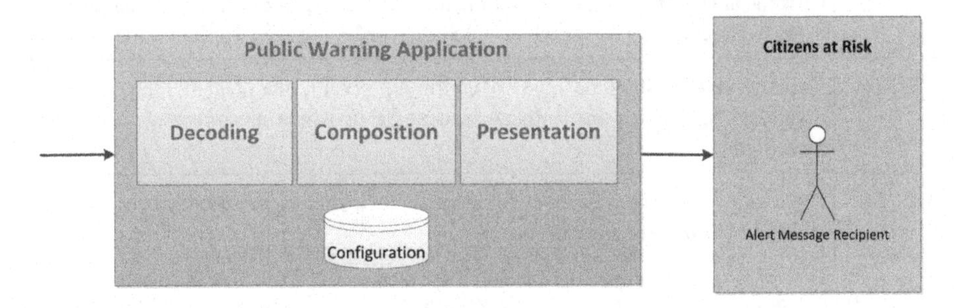

Figure 1.4. *Public warning application functional architecture*

1.5. Exemplary case: the Alert4All approach

Having introduced public warning systems as a whole, and in particular, public warning receivers and applications, this section will focus on public warning applications developed in the framework of the European Alert4All project [ALE 13], which have been further developed in the PHAROS project [PHA 16]. The Alert4All project developed an interdisciplinary alerting framework in order to improve the effectiveness of alerts and communication to the population at risk with a focus on a pan-European perspective. With this overall main objective, Alert4All developed an innovative multi-channel alerting system for the distribution of alert messages to the population at risk using a wide range of communication technologies and standard commercial receiver applications. The detailed list of communication technologies and the corresponding receiver devices that have been used in the Alert4All and PHAROS projects is presented in Table 1.4.

Communication technology	Receiver device
DVB-S/S2	HbbTV television
DVB-T2	HbbTV television
DVB-SH	Smartphone/siren
GNSS	GNSS receiver (simulated)
Mobile wireless networks (Cell broadcast)	Cell phone/smartphone

Table 1.4. *Communication technologies used in Alert4All and PHAROS*

Taking into account the technologies detailed in Table 1.4, several receiver applications have been developed in the framework of Alert4All. These applications have been designed and implemented in order to be compatible with the Common Alerting Protocol (CAP) [OAS 10] and also with the A4A protocol designed in the context of the Alert4All protocol [DEC 12a, DEC 12b]. Firstly, a CAP-compatible receiver application has been developed for HbbTV systems. This application should be installed as part of the firmware of the HbbTV-enabled television receiver. Secondly, a receiver application has been devised for the reception of alert messages using an available Internet connection (either via satellite, terrestrial or mobile wireless networks) in portable devices. This application was designed to be CAP and A4A compatible and to be installed in smartphones and GNSS receivers. Finally, the reception of alert messages over cell broadcast [ETS 10] and the trigger of sirens activated via satellite (DVB-SH) have also been implemented without the need for a dedicated receiver application. The different applications are described in the following sections.

1.5.1. *HbbTV-enabled receivers*

The implementation used to transmit alerting messages via HbbTV conforms to the HbbTV standard. Therefore, an HbbTV-compliant receiver should be able to interpret and show the alerts on the television set. The recipient only needs to have an HbbTV/DVB-compliant receiver to be able to receive and process the data carousel in which the alert message and the HbbTV application are sent.

HbbTV is mainly developed to merge broadcast and broadband together. However, the methodology that is used within the A4A project has the particularity of not assuming that the receiver is connected to the broadband connection. Therefore, all of the important information is transmitted via broadcast (DVB-T2/DVB-S/DVB-S2). One of the main reasons not to rely on the broadband connection is that this connection might not be available during emergency situations, and therefore it is possible that the alert message cannot be transmitted. In this case, the satellite communications provide robustness against most of the possible emergency situations.

Within the Alert4All project, a specific HbbTV mask has been implemented to display CAP-compatible messages on HbbTV-enabled television receivers. Figure 1.5 shows an example of an alert message displayed using this application together with its main elements. As can be observed, the application itself appears as a transparent overlay to be shown over the program that the recipients are watching. On the upper left corner, an identifier of the alert message issuer (the corresponding authority) is shown to allow recipients to identify the source of the message easily. The message to be shown is divided into three main parts: the headline, the risk or hazard description and the recommended protective actions (instruction). The headline is directly extracted from the "headline" CAP-field and is intended to give an overview of what happened in a nutshell. For good readability, it is recommended that the headline does not include more than 37 characters. Below the headline, there is the description extracted from the CAP-field description and, finally, the instruction, which are the recommended protective actions and the corresponding timing. For the sake of providing good readability, it is recommended that each section contains no more than 240 characters.

In the lower part of the application display, the three buttons, matched to the colors used in common TV remotes, provide additional features. The "exit" button closes the application and the TV goes back to the normal TV program; with the green button the user can choose between several languages (multi-language). The "More Info" button shows additional information, e.g. links and QR code to a special web page and also a map where the affected area is marked.

If more than two languages are transmitted within the original CAP alert, there are two solutions to handle this. In case of three different languages, there could be a fourth button to access this language (e.g. one button with "English" and another one with "Français"). If more than three languages are provided in the alert message, another solution has been defined, due to the limited colored buttons (i.e. four) on a standard remote control. In this case, there would be a button called, for example, more languages and by pressing it, there would be a pop-up list with all the additional languages that have been transmitted.

Figure 1.5. *HbbTV application and its main elements*

If an alert is triggered and the application pops up, the sound of the program will be muted. It shall also be feasible to play an alerting sound each time an alert is sent or updated. The application itself updates the variables every 10 s, so that any change in the alert will be shown on the screen after 10 s, at the latest.

With the HbbTV application, no location-based features have been implemented. Nevertheless, it is indeed possible to set up the postal code when installing a new TV or set-top box. If this is done, the HbbTV application can be configured so that only TVs with a particular postal code get the alert. This would require the implementation of this feature on the receiver manufacturer side. Finally, it should be noted that, although all the features considered for the use of HbbTV within the scope of Alert4All are compliant with the HbbTV standard, not all features are currently implemented in all the TV sets and set-top boxes available in the market.

1.5.2. *Portable receivers*

The second type of application developed in the context of the Alert4All project was a dedicated application to be installed in portable devices, such as smartphones, tablet PCs and GNSS receivers. These devices can receive alert messages either making use of an Internet connection (via satellite, terrestrial or wireless mobile networks) or via satellite navigation systems [BER 14]. In this case, since the receiver application must be manually installed by recipients, a download link is provided to make the application available to the population. The application has been developed for Android which offers more flexibility for development and deployment, while a prototype has also been developed for iOS. In order to overcome the issue of citizens not installing the application in their devices and therefore reducing the effectiveness of alert messages, the installation of public warning applications could be enforced so that all portable receiver manufacturers provide the devices with the application pre-installed.

The transmission of alert messages to be distributed to portable devices can be done using either the CAP or the A4A protocol [PÁR 16, DEC 12b]. When the device receives the message, it is parsed to retrieve the relevant information. Firstly, the application triggers a sound to alert the user and a window pops up (even if the device is in a standby mode) to show the alert message. In order to decode and compose the message, the application uses a decoder (based on the use of alerting libraries described in [PÁR 16] and [DEC 12a]) in the A4A case and a message composition engine. The message composition engine takes into account the corresponding alerting libraries for the language of the user, which include the syntax and grammar rules to generate a human-readable version of the alert message out of the decoded information items. If the user has not modified the settings, the application will also read the message using a text-to-speech engine.

Figure 1.6 shows the graphical user interface of the portable receiver application. The main interface is a message list from which the user has access to all received messages at all time. The messages are displayed in chronological order using the information included in the header. A range of colors help the user to identify the messages that are related to the same hazard or risk. New messages appear highlighted in orange to warn the user.

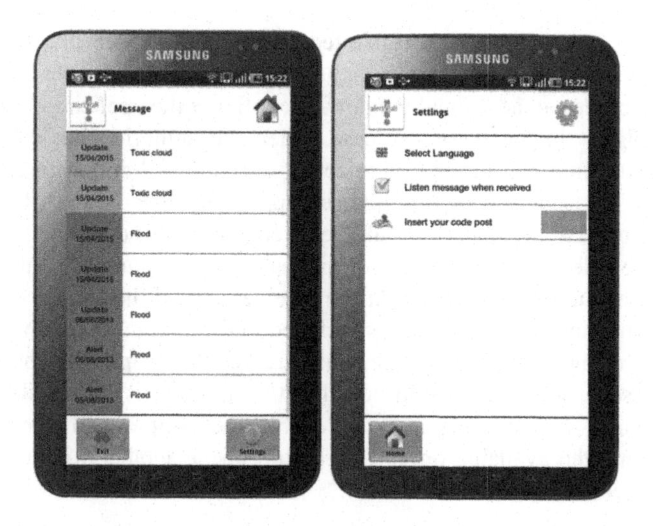

Figure 1.6. *Smartphone/GNSS receiver application user interface (left: message list; right: settings)*

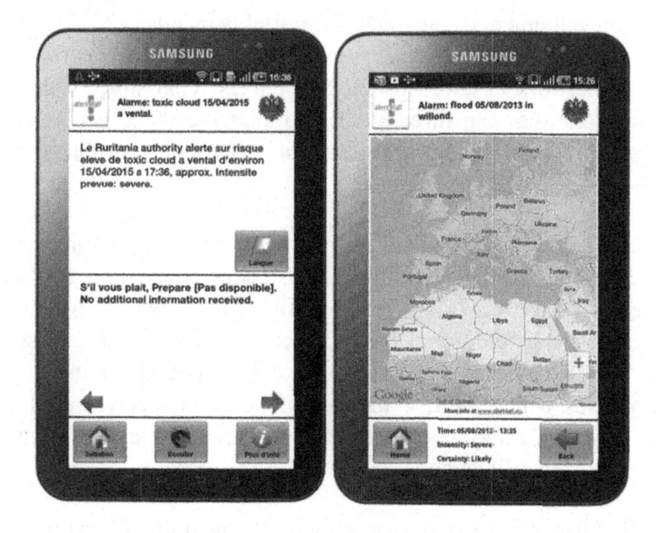

Figure 1.7. *Smartphone/GNSS receiver application user interface (left: alert message content; right: more info window)*

The alert message window displays the message as composed by the message composition engine (Figure 1.7 left). A button enables the user to listen to the message whenever it is requested. If the user wants to know more information, a button enables access to another window where the location of the incident can be seen on a map as well as the area affected by the event. More information about time, intensity and certainty is also included (see Figure 1.7 right). Finally, the application provides an additional window to configure the application settings (see Figure 1.6, right). In the particular case, when the authority that composed the message entered free text instead of selecting a predefined option to compose the message, thus not using the alerting libraries approach defined in [PÁR 16], several cases can be found. First of all, it should be noted that, in this case, the alert message issuer shall be offered the possibility of manually adding multi-lingual versions of the alert message to be transmitted. At the receiver side, if the language configured by the recipient is one of the languages included in the alert message, the message is shown to the recipient. If the language configured by the recipient is not included in the alert message, the user is informed that the message is not available for the chosen language. A button appears to enable the recipient to choose another language among the available ones. This button opens another window where the recipient can see the available translation.

1.6. Conclusions

The current development of personal and portable receiver devices that provide a wide range of communication interfaces (including the use of terrestrial, satellite and mobile wireless networks as well as GNSS) makes it possible to develop public warning applications that can directly receive alert messages generated by public warning systems without being processed or manipulated by any intermediate broadcaster.

The present chapter has provided an overview of public warning systems and applications, focusing on the communication technologies used and the receiver devices needed in each case. It has been highlighted that although the transmission of alert messages using traditional media, such as TV and radio, is still relevant, technological and behavioral constraints make it advisable to use multi-channel warning approaches which include the use of personal receiver devices, such as smartphones, tablet PCs and navigators to increase alert message penetration and therefore alert message efficiency.

The identification of relevant requirements to be fulfilled by public warning applications has allowed the identification of three main functionalities to be provided by public warning applications, namely decoding, composing the alert message and presenting it to the recipients.

Finally, the Alert4All/PHAROS approach has been used as an example of possible public warning applications which can be developed to distribute alert messages in a multi-channel approach, fulfilling the identified requirements.

1.7. Bibliography

[ALE 13] ALERT4ALL, "EU FP7 Project Alert4All", available at: http://www.alert4all.eu, 2013.

[BER 14] BERIOLI M., DE COLA T., RONGA L.S. *et al.*, "Satellite assisted delivery of alerts: a standardisation activity within ETSI", *Proceedings of 7th Advanced Satellite Multimedia Systems Conference and the 13th Signal Processing for Space Communications Workshop (ASMS/SPSC)*, Livorno, Italy, pp. 269–275, 8–10 September 2014.

[DEC 12a] DE COLA T., MULERO CHAVES J., PÁRRAGA NIEBLA C. *et al.*, "A novel protocol to transmit alert messages during crises over GNSS", *Proceedings of Ka Band and ICSSC Conference 2012*, Ottawa, Canada, 24–27, pp. 661–671, September 2012.

[DEC 12b] DE COLA T., MULERO CHAVES J., PÁRRAGA NIEBLA C., "Designing an efficient communications protocol to deliver alert messages to the population during crisis through GNSS", *Proceedings of 6th Advanced Satellite Multimedia Systems Conference and 12th Signal Processing for Space Communications Workshop*, Baiona, Spain, pp. 152–159, 5–7 September 2012.

[ETS 10] EUROPEAN TELECOMMUNICATIONS STANDARDS INSTITUTE (ETSI), Emergency Communications (EMTEL); European Public Warning System (EU-ALERT) using the Cell Broadcast Service, Technical Specification ETSI TS 102 900, Sophia Antipolis, France, 2010.

[INT 12] INTERNATIONAL FEDERATION OF RED CROSS AND RED CRESCENT SOCIETIES, "Community early warning systems: guiding principles", available at: http://www.ifrc.org/PageFiles/103323/1227800-IFRC-CEWS-Guiding-Principles-EN.pdf, Geneva, 2012.

[JOH 10] JOHNSON J., MITCHELL H., LAFORCE S. *et al.*, "Mobile emergency alerting made accessible", *International Journal of Emergency Management*, vol. 7, no. 1, pp. 88–99, 2010.

[LAN 10] LANGDON P., HOSKING I., "Inclusive wireless technology for emergency communications in the UK", *International Journal of Emergency Management*, vol. 7, no. 1, pp. 47–58, 2010.

[MUL 14] MULERO CHAVES J., PÁRRAGA NIEBLA C., "Effectiveness evaluation model for public alert systems", *Disaster Advances*, vol. 7, no. 8, pp. 1–11, August 2014.

[OAS 10] ORGANISATION FOR THE ADVANCEMENT OF STRUCTURED INFORMATION STANDARDS (OASIS), "Common Alerting Protocol (CAP) version 1.2", available at: http://docs.oasis-open.org/emergency/cap/v1.2/CAP-v1.2-os.pdf, Boston, 2010.

[PÁR 16] PÁRRAGA NIEBLA C., MULERO CHAVES J., DE COLA T., "Design aspects in multi-channel public warning systems", in *Wireless Public Safety Networks 2*, ISTE/Elsevier, pp. 227–261, 2016.

[PFE 13] PFEFFER R., SIEPE S., VOGEL B. *et al.*, "HbbTV a powerful asset to alert the population during crisis", *Proceedings of Networked and Electronic Media (NEM) Summit*, Nantes, France, pp. 58–62, 28–30 October 2013.

[PHA 16] PHAROS, "EU FP7 Project PHAROS", available at: http://www.pharos-fp7.eu/, 2016.

[SUL 10] SULLIVAN H., HÄKKINEN M., DEBLOIS K., "Communicating critical information using mobile phones to populations with special needs", *International Journal of Emergency Management*, vol. 7, no. 1, pp. 6–16, 2010.

[UNI 09] UNISDR, "Terminology on DDR", The United Nations Office for Disaster and Risk Reduction, Geneva, available at: www.unisdr.org/we/inform/terminology, 2009.

2

An Innovative and Economic Management of Earthquakes: Early Warnings and Situational Awareness in Real Time

2.1. Introduction

Current technologies, together with an increase in the importance of Emergency Management (EM) in recent years, have enabled significant improvements in the way in which natural disasters are dealt with, passing from a public action characterized by improvised measures and response capabilities of government to a management model characterized by the ability to prevent and adapt, leveraged by

---

Chapter written by Oscar Marcelo ZAMBRANO, Ana Maria ZAMBRANO, Manuel ESTEVE and Carlos PALAU.

the agility and effectiveness of emergency management systems (EMSs) and public safety networks (PSNs) [IZU 09].

Early Warning (EW) can be defined as a hazard warning mechanism. This requires processes and plans developed based on a situation awareness resulting from timely and effective control of the environment, as well as continuous and on-going interaction between all stakeholders. It refers to all processes related to the identification and evaluation of threats and risks present in the environment, and the corresponding event alert to stakeholders. An Early Warning System (EWS) is a set of resources and processes that allow carrying out EW, collaborating with the development of preparedness, prevention and protection capabilities and improving the resilience of whole community. Its main objective is to give the time to the affected social nucleus to mitigate the potential damages that may occur and protect the life and integrity of persons. EWSs are a fundamental part of government policies for public security and disaster relief (PPDR), and they have gained recognition in recent years because of their key role in reducing human and property losses at the moment of dealing with natural disasters.

Earthquakes are one of the most frequent and destructive natural disasters, costing billions of euros in facilities, properties and worse, human lives. About 80% of terrestrial earthquakes originate in the Pacific Circuit zone, an area of mountain and volcanic delimited by the Pacific Ocean. Earthquakes such as those in Ecuador in 1949 and 2016, Colombia in 1999, Peru in 2007 and Chile in 2010, are clear examples of its destructive power: all of them caused at least 8,000 deaths in just seconds. Despite this, some countries with high seismic risks like Ecuador [SEC 12], whose seismic data have been used to validate this architecture, do not yet have a proper EWS for earthquakes (EEWS). Therefore, the implementation of an agile, effective and functional EEWS, which does not involve a heavy investment and is compatible with the available resources of developing and permanent seismic risk countries like Ecuador, becomes a vital necessity.

This chapter presents an alternative for implementation of a low cost EEWS, based on the capabilities of sensing and communication of end-user smartphones (SPs) that are within the interest zone. It proposes a hierarchical three-level architecture: reliable, scalable and adaptable, which allows acquisition of real-time data of seismic events from a wireless sensor network formed by SPs, and if a seismic event of great magnitude occurs, immediately alerts all stakeholders.

The SPs are the central element of this architecture, which are polyvalent devices, and although their sensorization capability is lower than specific seismic devices (accelerographs), it is also true that it is economically impossible for a government to have hundreds of thousands of sensors that cover its entire territory. Current technological development has facilitated the massification of SPs; at the end of 2013, the number of mobile devices connected exceeded the number of people on the Earth, and by 2017, there will be nearly 1.4 mobile devices per person, which means over ten billion mobile devices connected, including machine-to-machine (M2M) modules that will exceed the world population at that time (7.6 billion) [CIS 13].

Implicit in architecture is the need to integrate and analyze the information from the sensor network, making it possible to obtain accurate and real situation awareness and identify seismic events independent of the user's usual movements. This information management starts from the data fusion model of the Joint Directors of Laboratories (JDL) [STE 99]. The hierarchical architecture, shown in Figure 2.2, begins with layer 1, which corresponds to a low-energy consumption application for SPs (e-quake), developed for Android OS (non-proprietary), that allows distinguishing between the user's usual movements and a real seismic event. Layer 2 represents a server with enough hardware and software resources to integrate and analyze all data from the SPs that have installed the application "*e-quake*". This server, named Intermediate Server (IS), performs a spatial and temporal analysis of samples from sensorization nodes (SPs), using a combination of mathematical methods that have not been considered so far in other investigations. The key contribution is at the real-time seismic events detection, capable of sending timely notifications to SP users at layer 1, making it possible to get extra time (minutes or hours) for better planning and decision-making that could be the difference between life and death for those affected. Finally, the third layer, the Control Center (CC), allows management during and after a damaging incident, collaborating with the aid centers and the network as a whole.

The life and care of persons is invaluable and it is our responsibility to know how to protect ourselves from harmful incidents. Having a support mechanism for preparedness, prevention and protection processes is as important as making good decisions during and after a disaster. This architecture also envisions a post-event management, where each SP helps the CC by sending additional information as comments, pictures and videos, collaborating with processes that achieve a comprehensive and real situational awareness, in line with disaster evolution and environment changes.

This work is organized as follows. First, previous and related projects in the area with their respective contribution are presented. In the next section, the proposed architecture is detailed and justified. Finally, the conclusions of our work are presented.

2.2. Motivation and previous works

When a seismic event occurs, the time available to safeguard the life and safety of persons is critical, can be the difference between life and death for those people affected, and is directly related to the ability to detect and report the disaster [KON 99]. This is the main objective of EWSs and this architecture: *obtain the required time to protect the lives and safety of persons*.

The Hyogo Framework for Action [UNI 06] emphasizes the need to identify and assess disaster risks and enhance EW. Local, national and international agencies related to the EM are making important efforts to incorporate EWSs as integral components in their PSNs and EMSs. The main features of an EWS are the detection and prediction of major natural disasters, in order to alert and report timely to the whole community of the occurrence of a damaging incident – in this case, a seismic event.

An EEWS bases its operation on major seismic events identification, and broadcasting warnings in the first seconds of their occurrence. This allows reporting the existence of the seismic event to the social nucleus at risk before the broader waves reach the site in question, allowing the authorities, first responders and people to take steps to mitigate their effects. For example, in Mexico 1995 (magnitude 8.0 earthquake), thousands of lives were saved because people were warned (50 s ahead) using an EEWS [ALL 11].

The proposed architecture has been developed in full using a non-proprietary platform, allowing the modification and customization of this solution to the needs of the population and region, which in case of a seismic event of great magnitude when many communication networks are often lost, makes it possible to find and offer an alternative for communications at response and recovery operations (WiFi, WiMAX, 3G or 4G). The SP application, *e-quake*, has been developed using free software (Java in Android), eliminating dependence for third parties or being subject to constant updates. The *e-quake* app should perform the sensorization and data collecting transparently to the users, always considering the battery charge optimization, and clearly identifying the seismic peaks of usual movements. The main contribution of this architecture and its difference with other works is on layer 2, which manipulates, integrates and processes the samples using the sliding window algorithm, Kruskal–Wallis ANOVA [KRU 04] and MQTT (Message Queue Telemetry Transport) [HUN 08] as the messaging protocol for information exchange and warning notification in real time. Another contribution is in the spatial analysis of samples by means of attenuation and distance Haversine equations [ROB 57], in order to find the optimal range (radio distance) that each IS must cover for the samples acceptance and to improve correlation and accuracy of processes. Finally, the addition of a third layer (CC), allows a post-event management, collaborating with aid centers and commanders to safeguard the life and safety of people. All these processes are properly explained in the following sections.

Most of the current studies on earthquakes are aimed at improving the detection and it is EWSs with different approaches, methods and technologies that make the difference between one and another proposal. Among these, we can mention the following:

– *QuakeCast* is a project focused on the detection of seismic events using a motionless 3-axis accelerometer and a personal computer. It determines the epicenter, intensity and location of an earthquake, and then generates the corresponding alert if necessary. It is oriented to warn first responders but its scope does not cover communities or citizens at risk. It has a high accuracy due to the evaluation of P and S waves [BOR 12] as seismic peaks sensing mechanism, but this accuracy is obtained at the cost of limitations on mobility and spatial coverage. By contrast, our system is aimed at the whole community, achieving greater reliability based on the increased user number, and minimizing the risk of errors in measurements due to device failure [CHA 11].

– i*Shake* is a system developed by Berkeley University (USA) that exploits the polyvalence of smartphones to measure the acceleration of a seismic event and predict the arrival of an earthquake. It is oriented to devices with Apple iOS operating systems, which greatly minimizes its scope, whereas the market trend is towards the use of non-proprietary software like Android [BIM 10]. This project is complemented by our work; taking advantage of the separation of the areas of spatial analysis does not take i*Shake* into consideration [DAS 13, REI 13].

– *Quakeshare Pro* is a proprietary system for detection and EW of earthquakes developed by CynSIS. It is based on the capabilities of sensing and communication of SPs, and their main difference with our proposal is in the third layer of our architecture, which adds functionality to the command and control of operations, and collaboration with other support and management centers [QUA 16].

– *Community Sensor Network (CSN)* was created by the California Institute of Technology (USA) and is currently one of the most complete projects. It collects and analyzes data from the SPs and implements a CLOUD application of earthquake management, using the virtual servers provided by the Google App Engine Architecture [GOO 15]. By relying on the infrastructure of a third party, it is obliged to comply with the conditions imposed by this mediator for use, as well as to adapt to the restrictions and functionalities on the platform [FAU 11].

2.3. Architecture

It is necessary to collect and validate the information from the wireless sensor network (layer 1) using a data integration process called *data fusion*, corresponding to the identification of a real earthquake based on acceleration and intensity of seismic events, and its representation within a consistent, accurate and useful data format. This architecture is an example of multi-sensor fusion, taking into account two premises:

– merging data from unreliable sources (reliability <0.5%), does not bring improvement;

– merging data from accurate sources (reliability <0.95%), does not bring significant improvement.

In conclusion, if the sensors are not sufficient in number or accurate enough, there is no sense in investing time and effort. From these two premises, the data fusion in this PSN is justified, since the SPs, although not even close to the accuracy of the specific seismic devices, maintain acceptable accuracy. This is also named *crowd sensing* [HUA 14, TAL 14] that nowadays is widely researched.

The hierarchical architecture is composed of three layers: the first is basically a network of sensor nodes SPs which gather data (x, y, z accelerations; latitude; longitude; date; hour; magnitude and intensity); the second one is composed of computers (servers) with enough hardware and software resources to enable receiving and sending data to all SPs which are in its coverage range and also to the third layer; layer 3 is composed of the CC, which has a certain, current and global vision of the situation, being responsible for the command and control of the whole architecture and interacting with the aid centers.

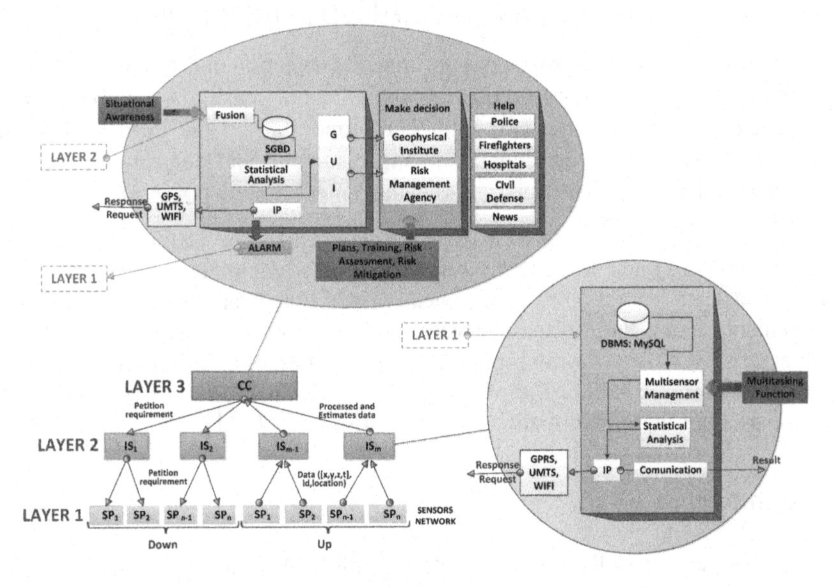

Figure 2.1. *Architecture*

The *Data Sources* are those devices that supply information; therefore, they represent each sensor in layer 1 within the hierarchical architecture; in other words, each SP is installed with the application “e-quake”. The e-quake application represents the *human-sensor interaction* that is capable of transmitting the data obtained to IS easily and efficiently. *DBMS* is the database management system where all the information is collected. A MySQL database has been used in this research.

For its part, the JDL model has five levels of processing, together with a data bus connection between the components, shown in Figures 2.1 and 2.2:

– *Level 0 – The fusion signal:* this makes it possible to reduce the amount of data and keep only the useful information. This is a background process which is performed on the SPs in layer 1;

– *Level 1 – Spatial and temporal validation of data*: this level corresponds to layer 2 in the hierarchical architecture, which has been developed using two programming languages: Java and Matlab. This application is situated in the IS, and is able to assess and validate all the information from SPs, and determine whether or not there has been an actual seismic event;

– *Level 2 – Situation assessment:* performed by the CC at layer 3, makes it possible to obtain a situation awareness of the environment;

– *Level 3 – Impact assessment:* identify risks, vulnerabilities and threats on the operational environment. The risk level of the seismic activity detected by level 2 and level 3 is assessed. These processes are the responsibility of CC;

– *Level 4 – Refinement of data fusion process:* data fusion process optimization and performance system control. It performs at all JDL levels.

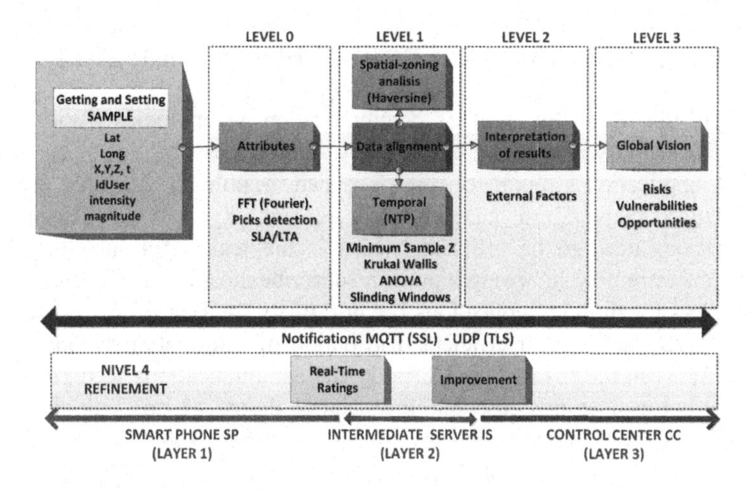

Figure 2.2. *Architecture*

It is important to note that the JDL levels do not relate to the proposed layers in the architecture. So, these two terms, layers and levels, must not be confused. The JDL model perfectly covers the three layers of architecture and enables compliance with an effective and agile data fusion.

The architecture has the following features:

– *Simple Maintenance:* the health of each sensor is the responsibility of the end user. Besides this, the user only takes a few seconds to install or update the application;

– *Adaptive and Scalable:* The more nodes the sensor network has, the greater the ISs' processing will be, but SPs will not perceive the change between a hundred, a thousand or even ten thousand SPs on the system. The architecture uses a lightweight communication protocol which requires a minimum consumption of resources (CPU, RAM, etc.), and allows new servers to be entered in order to collaborate with requests, notifications and processing for the servers that need it;

– *Timely:* by using real-time protocols and technologies for validation and expiration of data;

– *Secure:* first layer SPs send personal information but this could be intercepted during the communication and could be misused. TLS is proposed as a security mechanism to support data exchange over UDP, and MQTT is used for secure notification over SSL;

– *Standard:* by means of using standard protocols and technologies at every layer and COTS (commercial of-the-shelf) devices as consumer electronics networking hardware and SPs;

– *Synchronization:* NTP (Network Time Protocol) is used for clock synchronization between layers where CC is defined as Stratum 0 server NTP.

Between SPs and ISs, several communication technologies (WiFi, GPRS, WiMAX, 3G or 4G) can be used due to multi-interface capabilities of current SPs. From the application's point of view, it can seamlessly transfer from one communication network to another if a connectivity loss is detected. The communications need to be efficient, which is the reason for including MQTT protocol – an extremely lightweight publish/subscribe messaging transport protocol. It is also ideal for mobile applications because of its small size, low power usage, minimized data packets and efficient distribution of information to one or many receivers. Between the ISs and the CC node, the communications scenario becomes less tactical but even in this case data links can be destroyed, so satellite links may be an alternative. Communications are always real time, so that UDP [PAN 06] is used as a transport protocol in order to avoid retransmissions and connection-oriented processes.

2.3.1. *Smartphone application and acceleration processing*

The SP application must be simple, non-interfering, non-battery consuming; as well as a good helper during and after a seismic catastrophe in order to assist crisis managers and aid centers. Regarding the implementation issues, the Android operating system was chosen due to the following advantages:

– it has a strong position in the market guaranteeing a huge amount of potential users;

– it is free open source software, eliminating dependence on third parties and facilitates its customization;

– the application development is relatively easy;

– it has a properly manageable life cycle.

The application has been developed adapting to the capabilities and limitations of the Google API 2.2 platform ensuring that more than 95% of users using Android OS [DEV 15] would be able to install it smoothly. Furthermore, one of our goals is to have as many sensor nodes as possible.

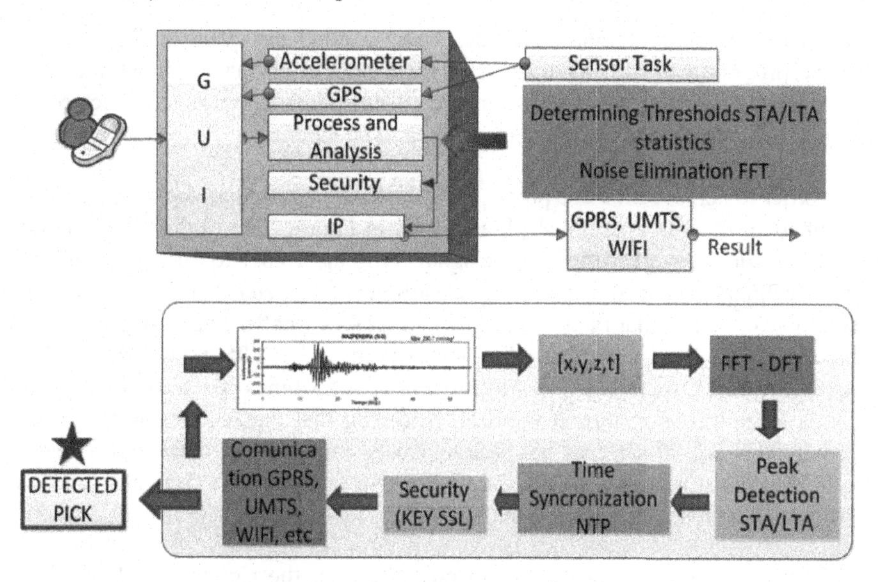

Figure 2.3. *Acceleration processing*

Figure 2.3 shows a block diagram that defines the process when the SP acts as an accelerograph, using its sensorization and network capacities. The SP application is named *e-quake*, which consists of a graphical user interface (GUI) for monitoring the sensors and determines whether a sudden and abrupt movement represents a possible seismic event. The application *e-quake* must fulfill the following characteristics for the success and acceptance of the system:

– *The user perception:* it is important that the application be made intuitive, easy handling and easy to understand; however, it is worth noting that most of the functions (sensoring and mathematical process) run in the background without requiring user interaction;

– *Transparent execution:* the application must control and quickly respond to failures that may occur, and continue running transparently to the user. The application must be made in such a way that does not disrupt the daily use of the SP; for the user it should be almost imperceptible that they have a new installed application;

– *Low power consumption:* this is critical, and has been one of the issues most considered in the design process; the application has been profiled thinking about the best way to solve energy consumption in the SP, otherwise users could discard *e-quake* due to its indiscriminate use of resources;

– *Free:* to get as many users as possible to ensure an adequate system performance;

– *Post-event support device:* besides being a good pre-event manager, the SP must be programmed to behave as an efficient post-event manager that allows the user to easily respond, receive information from other layers of architecture and obtain a real, current and comprehensive situation awareness.

In order to achieve all these points, with the SP being used as an accelerograph, a mathematical process is established, as shown in Figure 2.3 and discussed in detail below. In order to work with seismic signals ranging from 1 to 10 Hz (considering that the primary and secondary waves of earthquakes greater than 2 degrees on the Mercalli scale are in that range, and that earthquakes smaller than 6 degrees do not generally involve risk), the sampling interval must be 0.05 s (according to the Nyquist theorem); therefore it has to process data 25 times per second. Though it seems a large sampling rate, it is not, considering that the accelerometer can take approximately 35–40 samples per second. Peaks indicate that there was a seismic event. They must be handled with care because a false positive can cause chaos in the area and the loss of credibility in the EWS.

The time domain is not enough to properly treat the signal, so a background process collects acceleration samples, and by means of Discrete Fourier Transform (DFT), each coefficient is located in time and frequency [SHE 05]. This makes it possible to eliminate high frequencies that affect seismic signal peaks, considering that the discrimination of real seismic events is based on the evaluation of these signal peaks.

The STA/LTA algorithm [SHA 10] is then used to represent the signal-to-noise ratio (SNR) in order to define whether the peak at a short window (STA) versus a long window (LTA) exceeds a threshold. The STA/LTA algorithm is the most frequently used in the seismology field. An NTP client calls a server situated in CC to get an accurate clock time. Moreover, in order to handle user privacy (id, location), a security mechanism with TLS server key is used. Finally, a subscription/publication model with the MQTT M2M protocol provides the opportunity to select which ISs require an EW, as shown in Figure 2.4. MQTT also allows us to provide security with SSL, Quality of Services (QoS) and authentication with prefixes. When the SP receives a MQTT message from IS, the SP understands that a seismic event has occurred and it immediately notifies SP users with an alarm.

It is important to note that all detection processes are performed by SPs, and the data sent to the ISs will be only those which have met the constraints in order to be considered a real seismic event.

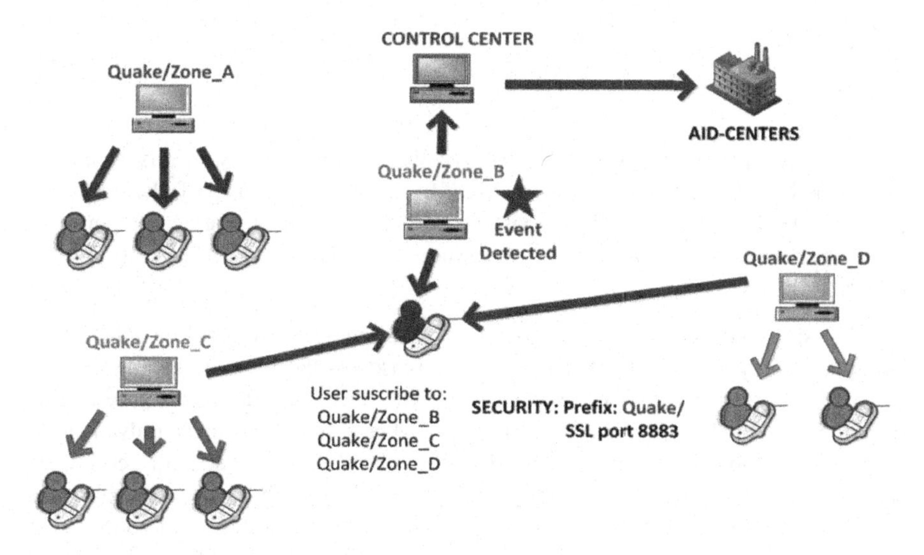

Figure 2.4. *MQTT notification*

2.3.2. *Server intermediate application*

For the second layer, the core software has been developed in JAVA and is currently able to run in 64-bit Windows and UNIX platforms as its main feature is the number crunching process to achieve accuracy and prompt responses. The IS always monitors the SP sent samples using different threads, as shown in Figure 2.5.

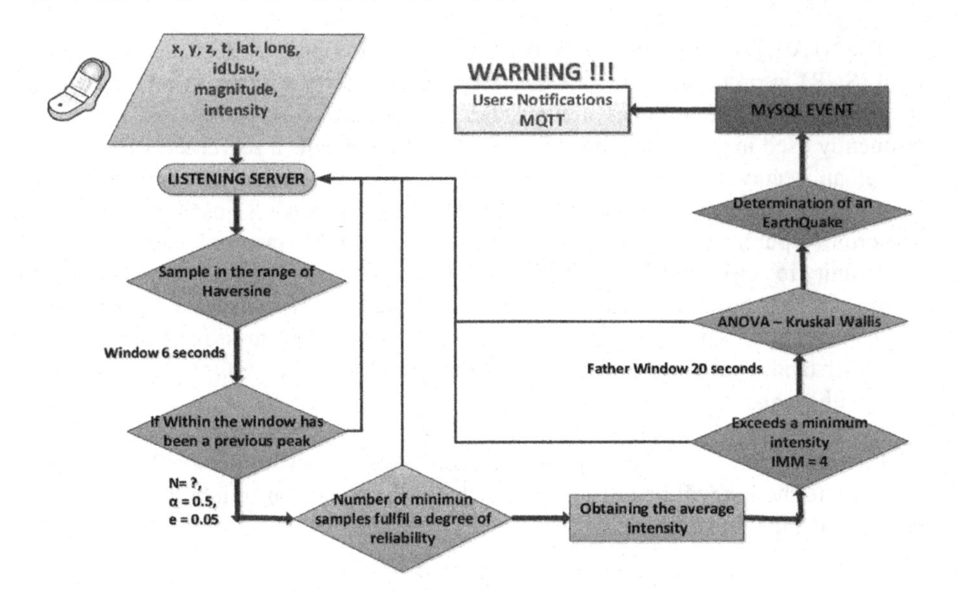

Figure 2.5. *Flow diagram IS*

It should keep its clock synchronized with the clock CC by means of a NTP server before entering them into a MySQL database because of the real-time scenario. Then, to determine the existence of an earthquake, the server performs some tests and validations in order to reduce false alarms which decrease system reliability:

– A spatial zoning analysis is a must; equation [2.1] defines the attenuation equations. They show that intensity ratio decreases as the distance increases, so if there were an event, the SPs in the area "A" would measure a greater acceleration than the SPs in the more distant zone "B". Therefore, an IS must only analyze samples that correspond to a single location or, in other words, spatially correlated samples. Using attenuation equations, we force an IS to only analyze samples that are within its range. Earthquake intensity varies in relation to these equations, and in turn they depend on soil types, speeds and other variables; so each geographic zone has its own equations. We work with one of Ecuador attenuation equations [BEA 10] as follows:

$$\log\left(A_{gals}\right) = 1.2474 + 0.3735 * M - 0.4383 * \ln\left(D+10\right) \quad [2.1]$$

$$A_{gals} = 1.2474 + 0.3735 * 5.5 - 0.4383 * \ln\left(D+10\right) \quad [2.2]$$

$$A_{max} = \sqrt{x^2 + y^2 + z^2} - g = \sqrt{19.6^2 + 19.6^2 + 19.6^2} - 9.8 = 24.14 \quad [2.3]$$

where A_{gals} is the acceleration in (cm/s^2), M is the magnitude and D is the hypocenter distance. We have set a magnitude M of 5.5 because according to the Mercalli Modified Intensity (IMM) of an earthquake, 5 is considered as moderate, and before this value it would not sense the warning. The maximum SP acceleration in three axes is used from equation [2.3] to obtain a balance between the number of samples and the effectiveness. If D is too small it might be the case that IS samples run out and if it is too large, the samples are uncorrelated, which means that the maximum distance to a hypocenter is equal to 35 km which corresponds to the epicenter distance [INS 11].

Samples that have a latitude and longitude that do not satisfy the Haversine function [ROB 57] between its location and the server's location saved in MySQL are discarded and will be considered by another IS that is closer.

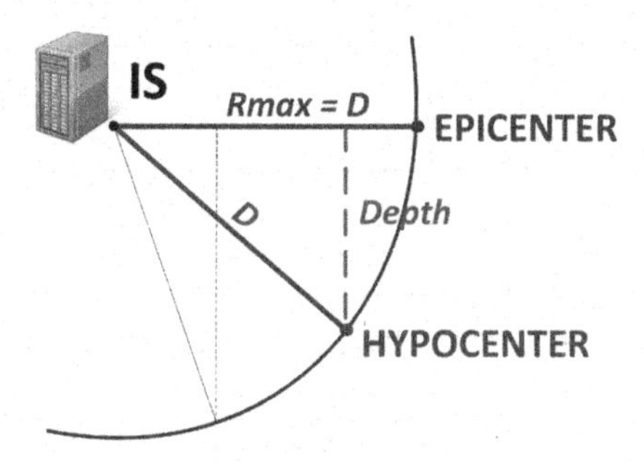

Figure 2.6. *Maximum hypocenter distance = epicenter distance*

– Minimum sample size refers to the number of active SPs (of all registered) that must detect a signal peak in order to assume that a seismic event has occurred.

The sample size will determine the accuracy and the margin of error in the detection process. If the sample size is too small to guarantee an acceptable reliability, is most likely that a seismic event did not occur, and this signal peak is due to sudden movements of some users who have coincided in time. The following equation defines the minimum sample size required, compared with the levels of reliability and tolerated errors:

$$n = \frac{N}{1 + \frac{e^2 (N-1)}{Z^2\, \sigma^2}} \qquad [2.4]$$

where N = total population, n = portion of population – number of users who have sent a seismic peak, e = permitted error (1%, 0.05%), Z = accepted reliability level (95%, 99%) and σ = number of typical errors.

According to this, the number of SPs that should detect a signal peak, or in other words, the number of samples that should receive an IS at a given time, must be equal or greater than the minimum sample size calculated to obtain a minimum reliability of 95% with a maximum margin of error of 5%. Table 2.1 shows some values obtained for the minimum sample required.

Users Number (N)	α = 0,01 = 99 %		α = 0,05 = 95 %	
	z = 2,57		z = 1,96	
	e = 0,05	e = 0,03	e = 0,05	e = 0,03
100	87	95	80	92
150	122	139	108	132
300	207	258	169	234
500	285	393	217	341
1000	398	647	278	516
2000	497	957	322	696
8000	610	1493	367	942
10000	620	1550	370	964
12000	626	1592	372	980
15000	633	1635	375	996
30000	646	1729	379	1030
50000	652	1770	381	1045
80000	655	1794	382	1053
100000	656	1802	383	1056
145000	657	1812	383	1059
150000	658	1813	383	1060
2000000	660	1833	384	1067
10300000	660	1834	384	1067

Table 2.1. *Minimum sample size*

Both SP and IS carry out validations to determine which SPs are alive (active) and which are not. First, SPs send beacons and are constantly monitoring the connectivity within PSN for reconnections, and second, the IS validates the last connection time and after a fixed time period (30 min) changes the SP state on database to inactive. These validations are important because we have considered a finite population which can change over time having due to SPs which do not have data location or network connectivity or have a dead battery, etc.

– Kruskal–Wallis or H test [KRU 04] is a non-parametric statistical test to probe whether a set of data comes from the same population (correlated) or not (uncorrelated), given that two groups are different when the variability between groups is greater than the variation within groups.

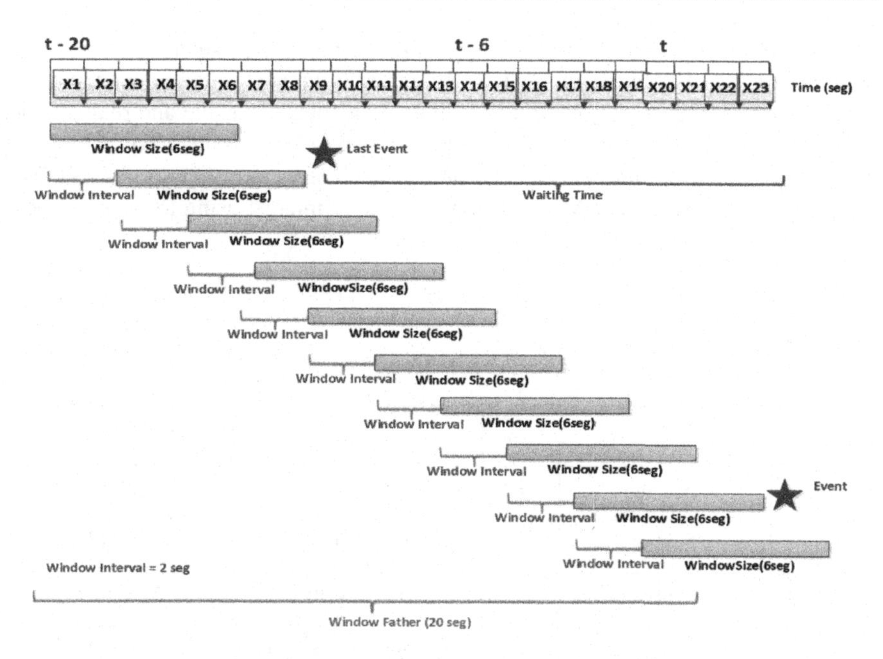

Figure 2.7. *Sliding window algorithm*

To use the Kruskal–Wallis test, a periodical sliding window algorithm was developed, as shown in Figure 2.7. The process repeats every 2 s, collecting the intensity of all samples stored in a database in a window of 6 s to process them. Each 20 s, which correspond to a Father Window, the intensities are taken again to test the correlation using Kruskal–Wallis and finally, trying to eliminate the risk of notifying the user of the same event or replicas, by validating that the time between the last and the present event has been at least 20 min.

Once an earthquake is detected, the user is informed immediately giving time to make a better decision. IS uses a MQTT server called Mosquitto [MOS 15] to publish to SPs who have subscribed to the topic whenever it detects an event.

2.3.3. *Control center*

The CC is considered the PSN central node, located at an EM agency headquarters acting as a command and control post, delivering accurate information to EM commanders about what is going on to help them make proper decision, thanks to the expensive and specific hardware (seismographs, accelerometers, inclinometer, etc.) that it has: for example in Ecuador, at the Geographic Institute at the National Polytechnic School [SER 15].

It allows us to extend from a pre-event to a post-event management, where first, each SP can help CC by sending additional information such as comments, pictures, videos, etc., helping in the process for achieving a comprehensive and real situation awareness of what is going on; and second, it can help users to make better post-event decisions, providing information about the closest aid centers, safer and faster routes, building a genuine EMS in the whole zone of interest using low-cost devices.

2.4. Results

Test equipment consists of the devices shown in Table 2.2.

Layer	*Type*	*Num.*	*Device*	*Operative system*
Layer 1	Sensor	2	Samsung Galaxy Ace II	Android (2.3.6)
Layer 1	Sensor	2	Samsung Galaxy S3	Android (4.1.2)
Layer 1	Sensor	3	HTC Desire	Android ()
Layer 1	Sensor	2	LG Optimus P700 – L7	Android (4.0.3)
Layer 1	Sensor	1	LG Optimus P470 – L5	Android (2.2.0)
Layer 2	Intermediate Server	1	Dell Inspiron Core i5	Microsoft Windows 7
Layer 3	Control Center	1	HP Pro Book Core i5	Microsoft Windows 7

Table 2.2. *Used devices by layers*

The SP application was tested simulating a seismic event by hitting a demonstration table such as in [SUZ 07], where the SPs rest. The results were obtained from this test (numbers obtained are an average of all SPs measures), as besides the unpredictable characteristics of an earthquake, it is impossible to try it in a completely real environment.

Although SPs do not have same quality accelerometers as specific devices, the application can gather movements as small as a message vibration (obviously, these movements are not sent to IS), verifying that the MEMS accelerometers have sufficient quality for this research as well as for other projects in the area as mentioned in section 2.2.

A relevant point to measure is the percentage of battery that the application consumes, as the user could stop using the application for this reason. At present, the application consumes a maximum of 10% of energy (42 h without charging) which if we compare with other applications is not too great, for example WhatsApp [WHA 15] consumes 13% on average, without the fact that our application could be the difference between being alive or not. This percentage could be decreased significantly by modifying the sampling rate, reconnection time or numbers of beacons. In addition, the application monitors this consumption often and if it detects a low level of battery (10%), it stops using the SP as sensor and just keeps running MQTT protocol for alerting the user. So, even if we lose a sensor, we can save a life just by keeping the SP functional.

A total of ten volunteers were asked to carry a SP while performing some usual activities such as walking, sitting or even running, in order to test whether our application can distinguish between real seismic peaks and these everyday movements by means of the STA/LTA algorithm. Here, the application functions with an accuracy of 85.1 %, or in other words, 15 in 100 times sent a message to the server when it was not a seismic event. This error percentage must be reduced by the respective IS, where all samples will be processed, significantly increasing accuracy and efficiency of the whole system.

The main goal of this work is providing extra time to the end user for better decision-making in the case of a seismic event. And we can probe that the architecture complies with the real-time system requirements: time that IS takes to process about 2,000 sent samples (2,000 active SP in the IS's range) by a program developed in Java to test the IS process (Haversine formula, Kruskal–Wallis and sliding window algorithm) is 1.2 s. Furthermore, there is a delay of less than 1 s between IS detects a seismic event and user is notified. These results in communication are a great progress, even outside WiFi environments such as 3G or 4G. This is due to the analyzed notification protocol, MQTT, that is currently widely used in research and projects related to the Internet of Things (IoT) [HUN 08].

2.5. Conclusions

A three-layered architecture has been developed satisfying the objectives of an EWS in terms of time, scalability, security and, adaptability in each layer, which increase its efficiency with the number of users in the system. In layer 1 (level 0 in the JDL model), a SP application was developed for Android OS, taking into account battery consumption and network reconnections, together with a notification mechanism MQTT that achieves a real-time system with a seismic detection mechanism that makes smartphones behave as accelerometers capable of differentiating between daily user movements (walking, running, talking, etc.) and real seismic events, in order to notify an IS that processes sent samples covering a calculated distance. This IS, corresponding to layer 2 (level 1 in the JDL model), was developed in Java and Matlab, and the main contribution of this research collects samples of a wireless sensor network formed by SPs in each geographic zone and processes them in order to identify real seismic events and notify the user, providing time to make a better decision that could save one or more lives. In the third layer (level 2 and level 3 in the JDL model), the Control Center, which is aware of the whole disaster scenario thanks to the information from the ISs, makes decisions in order to manage, in a more optimal and convenient manner than current best practices, all layers as well as aid centers.

A fundamental part, which has made it possible to obtain these results, has been the use of the JDL model. The JDL model, a data fusion model, has allowed the design of a safe and efficient architecture; it is scalable at each of its levels, allowing the correction of errors and optimally refining architecture. We can conclude that the architecture satisfies the real-time system requirements because it provides time that, although it may be only in the order of seconds (depends on the duration of a seismic event) or minutes, is vital and can save hundreds or potentially thousands of people's lives.

Finally, the results justify the methods used and the raised architecture, which could be implemented in a short time, with low-cost resources, leaving the efficiency to the number of clients in system, which we hope will become greater and greater. So, we have obtained a low cost and agile architecture that can be implemented anywhere, especially in places with high seismic risks such as Ecuador.

2.6. Bibliography

[ALL 11] ALLEN R. "Seconds before the Big One: Progress in Earthquake Alarms. Scientific American 2011", available at: http://www.scientificamerican.com/article.cfm?id=tsunami-seconds-before-the-big-one, 2011.

[BEA 10] BEAUVAL C., YEPES H., BAKUN W. *et al.*, "Historical earthquakes in the Sierra of Ecuador (1587–1996)", *Geophysical Journal International*, vol. 181, no. 3, pp. 613–633, 2010.

[BIM 10] BIMAL G., KHUSHBU S., Analysis of the emerging android market, MSc Thesis, San José State University, 2010.

[BOR 12] BORMANN P., ENGDAHL B., KIND R., "New Manual of Seismological Observatory Practice 2 (NMSOP2): Seismic Wave Propagation and Earth models, IASPEI", GFZ German Research Centre for Geosciences, Potsdam, pp. 1–105, 2012.

[CHA 11] CHANDY R., RITA A., QISTEIN S., "QuakeCast: Distributed Seismic Early Warning", *Caltech Undergraduate Research Journal Winter*, vol. 2009, 2011.

[CIS 13] CISCO Systems, "Cisco Visual Networking Index: Global Mobile Data Traffic Forecast Update, 2012–2017", available at: http://www.cisco.com/en/US/solutions/collateral/ns341/ns525/ns537/ns705/ns827/white_paper_c11-520862.pdf, 2013.

[COC 63] COCHRAN, W.G. *Sampling Techniques*, 2nd ed., John Wiley & Sons, New York, 1963.

[DAS 13] DASHTI S., BRAY J.D., REILLY J. *et al.*, iShake: The Reliability of Phones as Seismic Sensors, University of Colorado at Boulder, University of California at Berkeley, 2013.

[DEV 15] DEVELOPER ANDROID, available at: http://developer.android.com/about/dashboards/index.html, 2015.

[FAU 11] FAULKNER M., OLSON M., CHANDY R. *et al.*, "The next big one: detecting earthquakes and other rare events from community-based sensors", *ACM/IEEE International Conference on Information Processing in Sensor Networks IPSN'11*, p. 10, 2011.

[GOO 15] GOOGLE CLOUD, available at http://cloud.google.com/, 2015.

[HUA 14] HUANDONG M., DONG Z., PEIYAN Y., "Opportunities in mobile crowd sensing, *IEEE Communications Magazine*, vol. 1, no. 4, pp. 343–363, August 2014, 29–35.

[HUN 08] HUNKELER U., TRUONH H.L., STANDFORD-CLARK A, "MQTT-S A publish/subscribe protocol for Wireless Sensor Networks", *Communication Systems Software and Middleware and Workshops*, pp. 791–798, 2008.

[INS 11] INSTITUTO GEOFÍSICO DE LA ESCUELA POLITÉCNICA NACIONAL DEL ECUADOR, "Informe sísmico para el Ecuador", available at: http://www.igepn.edu.ec/images/collector/collection/informes-de-actividad-sismica/informe_ssmico_para_el_ao_2011_final.pdf, 2011.

[IZU 09] IZU B., MIGUEL J., "De la protección civil a la gestión de emergencias", *Revista Aragonesa de Administración Pública N.35*, Zaragoza, 2009.

[KON 99] KONTOGIANNIS T., KOSSIAVELOU Z., "Stress and team performance: principles and challenges for intelligent decision aids", *Safety Science*, vol.33, pp. 103–106, 1999.

[KRU 04] KRUSKAL W., WALLIS W.A., "Use of Ranks in One-Criterion Variance Analysis", *Journal of the American Statistical Association*, vol. 47, no. 260, pp. 583–621, 2004.

[MOS 15] MOSQUITTO. "MQTT Broker", available at: http://mosquitto.org/, 2015.

[PAN 06] PANAGIOTIS P., VASSILIS T., "On transport layer mechanisms for real-time QoS", *Journal of Mobile Multimedia*, vol. 1, no. 4, pp. 343–363, 2006.

[QUA 16] QUAKESHARE PRO, available at http://www.quakeshare.com/, 2016.

[REI 13] REILLY J., DASHTI S., ERVASTI M. *et al.*, "Mobile phones as seismologic sensors: automating data extraction for the iShake system", *IEEE Transactions on Automation Science and Engineering*, vol. 10, no. 2, pp. 242–251, 2013.

[ROB 57] ROBUSTO C., "The cosine-Haversine formula", *The American Mathematical Monthly*, vol. 64, no. 1, pp. 38–40, 1957.

[SEC 12] SECRETARÍA TÉCNICA DE GESTIÓN DE RIESGOS ECUADOR, "Guía para la incorporación de la variable de riesgo en la Gestión integral de nuevos proyectos de infraestructura", available at: http://www.gestionderiesgos.gob.ec/, pp. 99–108, 2012.

[SHA 10] SHARMA B., KLUMAR A., MURTHY V.M., "Evaluation of seismic events detection algorithms", *Journal Geological Society of India*, vol. 75, pp. 533–538, 2010.

[SHE 05] SHENG X., ZHANG Y., PHAM D. *et al.*, "Antileakage Fourier transform for seismic data regulation", *Geophysics*, vol. 70, no. 4, pp. 87–95, 2005.

[SER 15] SERVICIO NACIONAL DE SISMOLOGÍA Y VULCANOLOGÍA, available at: http://www.igepn.edu.ec/, 2015.

[STE 99] STEINBERG A., BOWMAN C., WHITE F., "Revisions to the JDL data fusion model", *Sensor Fusion: Architectures, Algorithms, and Applications*, vol. III, p. 430, 1999.

[SUZ 07] SUZUKI M., SARUWATARI S., KURATA N. *et al.*, "Demo abstract: a high-density earthquake monitoring system using wireless sensor networks", *Proceedings of the 5th International Conference on Embedded Networked Sensor Systems*, pp. 373–374, 2007.

[TAL 14] TALASILA M., CURTMOLA R., BORCEA C., *Mobile Crowd Sensing*, Department of Computer Science, New Jersey Institute of Technology, 2014.

[UNI 06] UNISDR, Early warning system, Report, Initiatives for disaster reduction, 2006.

[UNI 15] UNISDR, Global Assessment Report on Disaster Risk Reduction, Report, 2015.

[WHA 15] WHATSAPP, available at: http://www.whatsapp.com/, 2015.

3

Community Early Warning Systems

The community is represented as a network that interacts with society to expose physical and/or social impacts. Early warning is one of the many important tools that contribute to disaster prevention; we can talk about the term "disaster" as any negative event that impacts society. Therefore, a Community Early Warning System (CEWS) represents a set of capabilities required to achieve reliability and information security, minimize the information time, and manage time, human and monetary resources appropriately. Moreover, this is the reason why information technologies have greatly involved in this issue; to enlarge the benefits of a CEWS using new communication, protocols, sensors, etc. For example, smartphones today allow us to have all kinds of sensors within reach of our pocket, without

---

Chapter written by Ana Maria ZAMBRANO, Xavier CALDERÓN, Sebastian JARAMILLO, Oscar Marcelo ZAMBRANO, Manuel ESTEVE and Carlos PALAU.

communication of great features currently achieved; this certainly allows an efficient design and implant CEWS. Follows the main points that define and govern a CEWS, and in addition, some projects in different fields are presented demonstrating the efficiency of these systems.

3.1. Core early warning system components

An Early Warning System (EWS) allows harm and loss reduction with getting and disseminating warning information about hazards and vulnerabilities in a group of people who are considered at risk. Each word has an important meaning, for example, community involves a network of social interaction, early refers to prevention of any disaster or reduction of the potential harm or damage, warning means a message that announces danger and system puts all together. Therefore, a CEWS has four key elements [INT 12, ELL 12]:

1) risk knowledge;

2) monitoring;

3) response capability;

4) warning communication.

With this little introduction to a CEWS, it is important to keep in mind when a city becomes smart. A city is considered smart when it uses an infrastructure composed of communication technology, sensor and control devices, which collect data of what happens in the city and list all in an intelligent way, allowing for better resource utilization. The activities performed by their citizens are therefore faster and more effective.

With this knowledge about smart cities and CEWS, we can now analyze how these two ideas can be put together to obtain an effective CEWS based on smart city technology.

3.1.1. *Risk knowledge*

A warning system reduces the possibility of injury, death, damage and property loss from individuals and communities by immediately responding to the hazards. Therefore, it is important to understand the role that a smart city plays for disaster mitigation and decision-making for acquiring risk knowledge.

Smart cities have a proper communication and application infrastructure to support some characteristics about an effective early warning such as: decreasing the lead time and enhancing the accuracy of warnings, improving communication and disseminating warnings. Therefore, how can a smart city be integrated with a CEWS for risk knowledge? Firstly it is important to describe the four key elements of effective EWS. As described at the [EWC 06], knowledge of the risks faced technical monitoring and warning service, dissemination of meaningful warnings to those at risk, public awareness and preparedness to act.

Three of these four key elements can be supported by a smart city, because, the risk faced could be known by the research of hazards and vulnerabilities obtained by a smart city data infrastructure, monitoring and dissemination of the information could be achieved by smart city network.

In this context, the risk knowledge basically consists of knowledge about risks that include hazards, vulnerabilities and priorities at each level (global, regional, national or local).

If risk knowledge comes from communities considered to be at risk, it is possible to obtain very detailed knowledge; however, hazards and vulnerabilities can be obtained from geo-referenced maps, but using this method to obtain risk knowledge is not very reliable. As we know, smart city enables information collection about events that occur in the city and has a better utilization of resources; this feature can be used to get risk knowledge [HAN 12].

The smart city plays a role in this scenario with a specific application; it can provide necessary information to know where the hazards and vulnerabilities are by collecting information about historical risk (sensors). With proper research, the community planning could be better in different scenarios such as transportation, security, disasters, health, etc. [DEL 01]

One of the two principles for risk knowledge for EWS is important in a smart city context, and therefore, we have: "Although risk knowledge exercises may not lead to an early warning, all early warning must be founded on risk knowledge" [COW 14]. Risk knowledge must be founded on data, so it is difficult to acquire knowledge without information; the smart city supports this by providing accurate information for decision-making about hazards and vulnerabilities and how to control them effectively.

3.1.2. *Monitoring*

A sensor is a device that transforms environmental measurements into signals that can be understood and interpreted by a system; these measurements can be monitored, processed and stored by an application. Therefore, monitoring, as a component of EWS, should be considered as a key element, because it involves all of the information gatherings.

There are some technologies which are used to perform monitoring and information gathering in a smart city context. These are ZigBee, Dash7, 3G, LTE, RFID, NFC and Bluetooth. Through new protocols over the Internet, such as those used in the section, the utilization of anyone depends on the scenario that is applied [HAN 12].

The objective of a monitoring activity is recollecting data for risk knowledge. There are three crucial elements of monitoring: observation, measurement and prediction. Observation consists of an environmental behavior that anyone can see. Measurement is something expressed in numerical form, for example, the water level in a river or the intensity of an earthquake, etc. Prediction involves an analysis and is what is expected in the future based on this measurement. This probably triggers an action if said measurement reaches a (min, max) level.

Monitoring is considered the component that is most related to science and technology. In this aspect, EWS and smart city can be effectively integrated, because smart city makes all in an intelligent way by integrating communication and technology through monitoring and control devices for an efficient utilization of resources, as achieved when using smartphones as sensors. The core of a smart city infrastructure is detecting. With this feature, a smart city can monitor itself and take care of public infrastructures, based on the data collected by the sensors. To reduce costs, a smart city has been forced to implement real-time monitoring, which allows to reduce scheduled inspections significantly [HAN 12].

Sensors are the main component in a control system, because it collects data and sends to a central information system, where data are processed and stored for intelligent decisions. There are detection platforms that could be implemented in a monitoring system to perform sensing tasks, for example:

– Internet of Things (IoT);

– Cloud of Things (CoT);

– Sensor web;

– Sensor networks.

The Internet of things connects everything to the Internet, and the Cloud of Things is an IoT framework for service providers. These concepts are applied in a smart city because cities need an infrastructure to monitor and control their behavior, hence it is important to select the best option to achieve good monitoring and information gathering. For this, there is an instrument concept called Sensor web, which, in simple terms, consists of sensor pods strategically located and intra-communicated to monitor environments. Sensor webs have a unique feature that distributed sensors and sensor networks do not. This feature consists that information gathered by one sensor is used and shared by other sensors. On the other hand, in a sensor network, each sensor recollects data and does not have an impact on another sensor. Therefore, this feature of sensor webs allows modifying behavior based on collected data [DEL 01].

Smart cities have many applications, for example water distribution systems, electricity distribution system, smart buildings and homes, environmental monitoring, surveillance, health care and crowdsourcing. An important application with smart cities and their smart buildings in a CEWS context is an evacuation. For example, in a conventional evacuation scenario as for an earthquake, the system assumes that people inside know the building's exit routes and an alarm triggers when an emergency occurs.

The best way to get this, in any field, is to achieve the best early warning time. The anticipation of an "impact" allows better decision making in a matter of seconds, such as shutting down industrial machinery, stopping a train and exiting a burning building.

3.1.3. *Response capability*

Nowadays, the fast-growing urbanization continues exerting influence on substantial issues such as urban mobility and capacity to respond to disasters, among others, which lead to the need to cope with these challenges that include topics such as the impact of the environment or the ability to anticipate and react quickly to any disaster in cities.

That is why, on having realized the development of an EWS, it is necessary to keep in mind that, in a society, there are diverse groups, and during a natural disaster present different vulnerabilities like culture, disabilities and age, amongst others that play an important role when actively coping with a disaster, to prepare and to be able to have a correct answer to it.

Thanks to the relationship between systems, the information and communication technologies play a very important role in bringing the biggest profit to this "network" controlled by citizens. Therefore, a smart city uses technology that allows us to perform the transformation of the basic systems and their optimization, spurring them to innovation, and being a key factor for competitiveness and economic growth.

3.1.4. *Warning communication*

When a natural disaster occurs, the authorities responsible for carrying out an emergency plan should start by answering several questions, for example how will the alerts be given to the people who are in danger? Or, does the population have the ability to know and understand the warning that is given to them at a critical moment?

It is in this way that the authorities displayed on the communication of a warning, as a fundamental part when deciding on EWS. This decision is directly related to the information obtained from monitoring conducted in the most affected areas. To realize that the message is retrieved of a form that is clear and efficient, there are three main actors [MER 10]:

– Author: responsible for establishing and setting up the message that will be subsequently broadcast;

– Mediator: the first receiver of the message that must maintain the message original or can be modified in case of being necessary for an effective broadcasting;

– Receiver: a population that is responsible for understanding the message of warning delivered.

Taking the aforementioned parameters into account, we emphasise Information and Communication Technologies (ICT) because they allow us to have an efficient and effective way to obtain various data to prevent as is, in this case, a possible natural disaster. By using the ICTs, it is possible to integrate an EWS into any technical service, such as transportation, meteorology, seismology and health, which can generate information in the first moments of traffic jams, car crashes, floods [JUN 15], earthquakes or volcanic eruptions. These natural disasters can be prevented using several tools, such as:

1) system monitoring: to monitor first affected areas or future disaster areas;

2) the sensor of flood for the urban area;

3) the sensor of opening and closing of river gates;

4) monitoring of transit;

5) monitoring of landslide;

6) the collapse of river bed alarm.

To alert the population about potential natural disasters instantly, an intelligent alert system can be developed by bringing together a set of sensor networks, alarms and monitoring devices, which complements a smart city [FUR 10].

Per the mode of operation adopted, we can have [XIN 15]:

Modality	Community EWS	EWS centralized or instrumental	EWS mixed
Characteristic	Active participation of the population in the monitoring, using handmade instruments	Is characterized by using the automated monitoring	They use automated instruments and local monitoring
Advantage	It involves an active participation of the population The instruments used are of lower cost The trend is to lower use of this modality	A high technological component which will be effective The trend is greater use of this modality	It allows to be provided with technological advances, and too with the local monitoring The tendency is little use of this form
Disadvantage	The instruments used may not be accurate in critical moments	The technology used is of greater cost Demand for a good level of capacity of the operators	The union of these two modalities involves higher costs

Table 3.1. *Operation modes*

Technological advances have improved the quality and response time of warnings of dangers achieving inform technologies of surveillance and early warning. In Table 3.2, some technologies are observed:

Technology	Characteristics
Forecasting and modeling	They used data from monitoring including values of temperature, precipitation and climate models equipped with modern technology
Remote sensing and geographical information system	Allows us to give early warnings on food security before the end of the season
Satellite communication	Help reduce the time between the collection of data and the alert
Mobile technology	Used to communicate the alert and coordinate the activities of preparation, especially SMS alerts for mass mailings
ICT for multiple sources	Were extensively used in the Haiti earthquake 2010, thereby acknowledging that it could help in activities prior to the disaster by identifying potential risks
Cartographers crisis	They can provide information in real time about the impending crisis in times of uncertainty and confusion

Table 3.2. *Technological advances*

3.2. Time scenarios for EWS [EST 15]

3.2.1. *Real-time systems*

Real-time systems (RTS) have been developed and have grown in demand in the market especially in industrial environments.

RTS are considered to be systems whose behavior depends on the time elapsed, since they start processing data entries until the outputs are known. One of the most important features is that the response time of these systems must be predictable and limited. RTS can be classified into several forms, according to the fulfillment of the temporary restrictions:

– critical real-time systems;

– non-critical real-time systems.

3.2.1.1. *Critical real-time systems*

Critical RTS are systems in which the non-compliance with the temporary restrictions can have consequences insurmountable for the system itself and its users. These systems in critical real time are also called hard real-time systems. Some examples of these are earthquake alerts, military tactical training, kidnapping and rescue systems, etc.

3.2.1.2. *Non-critical real-time systems*

Non-critical RTS are systems in which failure to comply with time restrictions implies a loss of functionality or performance of the system. These non-critical RTS are also called soft real time. The multimedia systems (audio, images and video), are a typical example of these.

Another classification of real-time systems is given depending on the number of processors that are used for the execution of their tasks. Some types of RTS are:

– Conventional real-time systems: systems in which their tasks are running a single processor;

– Distributed real-time systems: systems in which not all tasks are running on the same processor. In addition, these systems can be strongly coupled when processors are working with the same operating system. Some of the features of these systems of real-time distribution are: systems concurrent, reagents systems, systems that operate in environments hard (high temperatures, noise, etc.), and systems that are inherently reliable and tolerant to failures.

3.2.2. *Management of the time in the distributed real-time systems [COU 11, JOA 01, TAN 08]*

As these systems take time as the main magnitude, the administration of the same one is important. For the management of these systems can treat the time as:

1) discrete magnitude;

2) common reference;

3) base of correction in the system;

4) base for the communication and synchronization.

Time management will focus on the synchronization of clocks in RTS. There are a series of algorithms of consensus distributed, whose objective is to establish a time base common among all the processes on a real-time system distributed.

The devices in a distributed RTS have clocks that provide the synchronization of the system processes; the difference in the time displayed in the system clocks is called bias (skew).

That is why it is important to have clocks that mark the time perfectly. The oscillation frequency that generates the clocks can vary by the environment and the same clocks may have a different oscillation frequency by the temperature. Owing to these characteristics, the clocks have a deviation or drift (ρ), as shown in Figure 3.1.

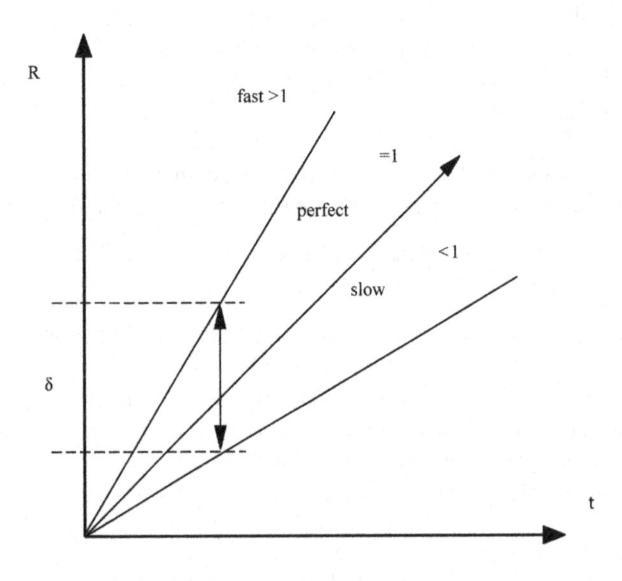

Figure 3.1. *Clock deviation*

Figure 3.1 shows that if its deviation (derives) is zero, the clock will be perfect, if its deviation is greater than 1, the clock will be faster and if the deviation is less than 1, the clock will be slow regarding this reference.

To consider a local time (for example, of a computer) like not erroneous, it is necessary to fulfill the following restriction:

$$1-\rho \leq \frac{dR}{dt} \leq 1+\rho \qquad [3.1]$$

$$-\frac{\rho}{2} \leq \frac{dR}{dt} - 1 \leq \frac{\rho}{2} \qquad [3.2]$$

It is known that the clocks tend to be perfect depending on the material with which it is manufactured, for example:

Clock	ρ
Malo	10–5
Quartz	10–6
Precision Quartz	10–7 or 10–8
Atomic	10–13

Table 3.3. *Clocks*

3.3. Core early warning system components using smartphones

The increased use of mobile phones worldwide, which have better processing capabilities, storage, speed data transmission and connection to the Internet using heterogeneous networks, has opened up a range of possibilities for the use of these devices in traffic monitoring applications, personal security and sending alerts in the event of disaster. Smartphones are especially useful for the latter type of applications, such as [THE 06, IBM 16, SMA 16, MIN 16], as smartphones offer this kind of applications four important characteristics: they are the device of information exchange used by the general public [BIL 16], are permanently connected to the Internet and cellular networks, have the ability to display video content, and have sensors to record its position, vibration, sound and images. These capabilities make them especially useful to broadcast alerts and collect information in a consistent, specific, clear and direct way to the population at risk before, during and after a disaster.

3.3.1. *Technology progress*

It is necessary to recognize the importance of wireless communication networks, as these systems have progressed from voice services to providing users with access to Internet, reaching 3.24 billion people in 2015 (44% of the world population) connected via mobile Internet [BBC 16], promoting the development of economic and social opportunities.

This development in the wireless communications infrastructure, specifically in the cellular system is exploited by EWS for the transmission of alerts through the Internet or via SMS messages to reach the population at risk with a broad range of information coordinated and published by the disaster management systems

[THE 06] and [SMA 16], decreasing the reaction time of the population to take actions to keep themselves safe.

3.3.2. *Efficiency to warn the population*

The information on the level of penetration of mobile phone use shows that the people around the world are adopting smartphones as the new way to stay connected with each other. For example 47% of US households have cellular service only, and 42% have landlines and cellular access [THE 16], which shows the level of influence that mobile telephony has on households. On the other hand, the data in [MQT 16] show that smart devices will comprise 66% of all mobile devices by 2020 [CIS 15]. For these reasons, the use of smartphones for EWS systems is a good solution to dispense important information to the population in masse about what to do in case of a disaster.

The use of smartphones also opens the possibility of using applications on low-bandwidth cellular systems, allowing them to alert the population that does not have electricity, and submit coordinated messages from different agencies that have the responsibility to issue notifications during emergencies. In addition, these applications can contain pre-saved information and tips about what we should do when faced with different types of risks during an emergency.

3.3.3. *Data collection*

Sometimes when a natural disaster occurs, power networks stop working for hours or days, depending on the situation. In which case the voice service in cellular networks and public telephony collapses rapidly due to the number of users wishing to make use of these services. In this instance the only way to transmit messages to the population is through devices operating by batteries, such as portable radios. However, this communication method only allows transmitting information to the population in one medium, and leaves no opportunity to send data about their current, specific needs and locations.

An alternative to this situation appears with one of the new theories for smart sensor networks, for example [MQT 15], which is the use of the information provided by users through a mobile application to collect information first-hand via the Internet, and get an overview of the damage, lack of basic resources, sanitary conditions, etc. through the transfer of small amounts of data (<1 KB per message [GAN 11]) using protocols such as XMPP or SIP. All the information gathered will be analyzed by government organizations responsible for coordinating the actions on the amount of resources to be sent to each affected population.

For example, in [MIN 16], there are recommendations about requirements for alert notifications to include people with disabilities [ASK 16]; these are made based on existing authoritative work of organizations, such as OASIS, W3C, NDIS, etc. Among the most important recommendations are: to provide detailed messages via text, audio, images, in different languages and forms of presentation that fit the display device; all these requirements can be met using smartphones, as these allow people with disabilities to be alerted promptly through sound alerts/audio, vibration and written/symbolic messages, enabling them to immediately understand an alert message that affects the area they are in.

In conclusion, the EWSs must be able to reach the people who live in an area affected by a natural or human disaster in a clear, consistent, timely and direct manner, which sometimes the regular broadcasting services such as radio and television are not able to provide [GSM 16, ANS 14]. This is why it is important to realize the integration of a new form to reach people, by using mobile applications in smartphones, capable of centralizing alerts issued by official bodies on all types of threats that exist against the life and health of people, providing a simple, rapid and inclusive form of interchange of information that users can use to stay safe, which comply with the recommendations of specialized agencies made in these cases [FAZ 12].

3.4. A smart city using smartphones into CEWS

The main idea is to take advantage of the growing trend of worldwide diffusion of smartphones, where almost each person has one or more of these electronic devices, which initially were created exclusively for telephony. However, currently these have some sensors, GPS, internet access and different types of communications [HER 11]. This leads to a host of applications being put into production.

So far, it has been shown that smartphones have been part of the great evolution of smart cities, thus providing their processing capacity, storage and different types of communication (Bluetooth, WiFi, 3G, 4G, NFC) to solve everyday problems. These devices, commercial off-the-shelf (COTS) and their communication with the Internet, deliver a number of new facilities and advantages that were previously defined in point 1.3, emphasizing real-time notifications, interaction with heterogeneous and low-power consumption sensors along with multipoint communication.

The CEWS that take advantage of smartphones are booming [JOH 10]. It can be applied in many areas of interest, such as transportation, security and natural disasters, amongst others. Each proposal of this type should combine the information from such sensors. This process is called "data fusion", which corresponds to a real object, in this case, called "seismic intensity", within a consistent representation, accurate and useful. All next projects are a clear example of a multi-sensor data fusion, based on two premises:

– merging data from unreliable sources (reliability <0.5) does not provide any improvement;

– merging accurate data sources (reliability >0.95) does not provide any significant improvement. Therefore, if our sensors are very accurate, it does not make sense to invest that effort. However, if this were the case, a fusion of information would improve the outcome.

Therefore, smartphones whose sensors have an acceptable accuracy are used in all types of projects as an efficient solution [TOR 13]. Therefore, the more the sensors, the less the error, as shown in Figure 3.2.

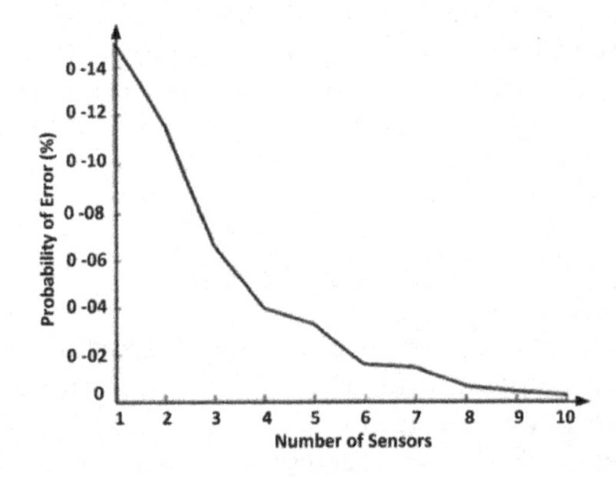

Figure 3.2. *Number of sensors versus error probability*

To give a better view of everything mentioned, three areas will be reviewed in which CEWS and smartphones have been an efficient solution:

3.4.1. *Natural disasters*

An early warning management for emergencies has gained notoriety among the population because of the crucial role it plays in reducing the losses of facilities, goods, properties and, more importantly, human lives. However, this requires an adequate preventive emergency system, which is not usually deployed in developing countries [JAY 06]. As an example, seismic activity is one of the most frequent and destructive natural disasters, and about 80% of terrestrial earthquakes originated in

Pacific-circuit zone: recent earthquakes such as the one in Ecuador, in April 2016, with more than 2,145 aftershocks which caused more than 1,000 deaths in seconds [IGE 16].

It is important to note that an earthquake can never be predicted; therefore, the best choice is "prevention"; and it is for this reason that universities and research centers have contributed efficient and economical solutions. For example, a seismograph can have a cost of tens of thousands of euros. For this reason, it is economically impossible for a developing nation, like Ecuador, to cover all its territory with these sensors. This is where the CEWS, that use smartphones, are the solution.

Currently, it is possible to find a lot of applications that aim to solve seismology, for example IShake [ERV 11], CSN [FAU 11] or E-quake [ZAM 14]; probably, these projects are the most representative in this field. CSN is developed by the Caltech Institute having a large number of users within his network, performs its analysis on Google Cloud Architecture and, later, issued warning alerts to citizens. On the other hand, IShake, which was developed by the University of Berkley, determines a next earthquake using only devices with Apple OS and processes it on an external server, and finally, E-quake, one innovatively and an economic proposal that manages to anticipate the arrival of an earthquake through the CEWS.

Each one has a different design and development providing advantages and disadvantages relative to one another. Table 3.4 provides this information:

Characteristics project	I-Shake	QuakeCast	E-quake
Smartphones as sensors	Yes	Yes	Yes
EWS	Yes	Yes	Yes
Community support (CEWS)	No	No	Yes
high rate of false alarms	Medium	Medium	Medium
Battery consumption	25 h	25 h	48 h
Operative system	iOS	Android	Any OS. Heterogeneity

Table 3.4. *Comparison of disaster projects*

From the same chart, we can deduce some points. I-Shake presents the limitation of including only iOS devices; therefore, much of the public would be unable to contribute. On the other hand, QuakeCast [CHA 09] uses a P and S wave's detection method, achieving high detection effectiveness in epicenters and magnitudes;

however, it is a requirement to keep the sensors at rest (no motion). Therefore, this project could not be used in a real and community environment. QuakeCast could be used in a no community environment where smartphones are exclusively sensors and nothing else.

E-quake is a proposal that includes IoT platforms and protocols. On the one hand, Sensor Web Enablement (SWE) [OPE 16] is able to analyze samples of any type of sensor, regardless of their characteristics. In other words, it offers heterogeneity; and second, allowing MQTT [MQT 15] sends notifications in real time.

It is implied that a process of detection within smartphones is necessary. Making a small comparison between these projects, we can see that the most efficient is E-quake, by taking into account two fundamental points: detection rate and energy consumption for this mathematical process. It is important to emphasize that the energy consumption is the reason that people do not continue in the CEWS process, and on the contrary, uninstall the application.

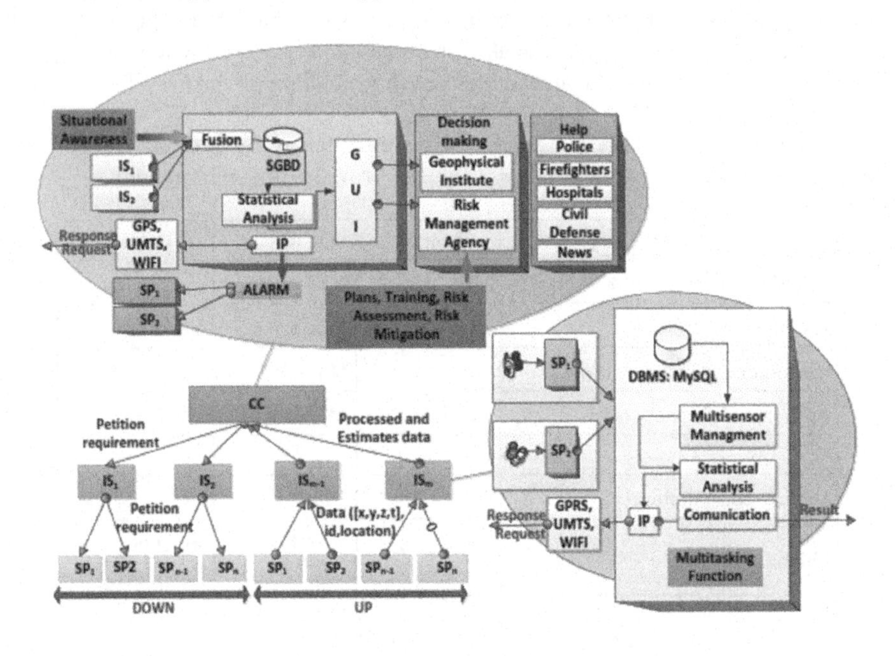

Figure 3.3. *E-quake architecture*

The accelerograph network developed in this paper is based on a three-layered hierarchical architecture for EEWS, as shown Figure 3.1. On Layer-1, SPs are used as processing units and they send samples to the intermediate server (IS), corresponding to Layer-2 as soon as SP detects a seismic-peak after overcoming a process that has been specifically designed. Each IS decides whether there was a seismic event or not, and immediately notifies their own users whilst at the same time, communicates the incident to the control center (CC) – the third layer. The data gathered from the sensors are inserted in an SOS (SWE) into the IS. The CC aggregates different applications that make decisions based on the information available in each SOS. The three layers and the SOS will be integrated in a scalable manner (one or more SPs, one or more ISs) until completing the system in order to verify that they interact properly, cover the functionalities and conform to the requirements and, furthermore, contribute to non-functional requirements, such as agile and easy portability, simple maintenance, to ensure the integrity, confidentiality and availability of information (security) in the architecture, reduce the cost of locating errors and provide an indispensable, economic, huge sensor network.

3.4.2. *Safe*

One of the biggest problems today in every society is the insecurity, and the more vulnerable or helpless the person, more likely to become a victim, for example, children, elderly or people with disabilities.

Although a large percentage of kidnapped people are found alive, it is also true that the vast majority are rescued from the seventh day (rescue time too long), which could represent irreparable damage to the person. Therefore, it is expected that this rescue time will be reduced from days to hours, and in the best scenario, this rescue process takes no more than tens of minutes. According to the International Center for Missing and Exploited Children (CINDE), more than 1.8 million children are victims of sexual exploitation globally. According to the annual report (2013) presented in Athens, Greece, 29 European countries had more than 630,000 calls denouncing child disappearances, serving 5,065 cases. This report also argues that each year more than 250,000 children go missing in Europe [MIN 16]. In the United States, more than 700 children are kidnapped a day and more than 250,000 a year; in other words, every 40 seconds a child is lost or kidnapped [BBC 16]. A comprehensive study of CINDE and UNICEF on the disappearance of children in Central America, in 2011, found "a serious problem of child trafficking by the existence of 'blind spots' at borders and increased illegal migration across the region" [BBC 16]. Although, fortunately, child figures are declining in Latin America, the current problem is the trafficking of girls and young women aged 11–22 years, as stated by the founder of an association of stolen and missing children [BBC 16].

There are several proposals in the field of safety, for example, the mobile application "I'm getting kidnapped" [FOR 16], which was developed by an engineer in Brooklyn, just in time for the World which could help make you feel a bit safer, Cup. So, how does an app help in a very physical and scary situation like a kidnapping? This application works by a pre-configured SMS alert, together with a pin location in Google Maps posts. For example, "Hello, if you're receiving this, Alex is in danger. Here is his last known location and please alert Authorities". It works almost in a real-time system, fulfilling the requirements involved, and faster than sending a deliberate message asking for help text. It is plausibly faster than calling 911 or the local police, who then have to know the exact situation and figure out where you are. However, it is also true that the victim should be able to open and send the pre-made message; it is necessary to keep the mobile signal to send the SMS, and its biggest limitation is the fact of using only the operating system Android, because Apple does not allow an application that is not designed for sending the SMS.

On the other hand, the "SOS app" developed by NowForce [NOW 16], in Israel, to call for help when they are in trouble, has already been in use for several years, giving very good results in the country. NowForce is well known in the emergency services community. The company's apps are used by fire and rescue; for example, the app is used by fire officials in Boone County, Missouri, to keep track of emergencies in the 500 square mile area they are responsible for. The app immediately alerts volunteers in the area when a 911 emergency call comes in reporting a fire or kidnapping. With the app, response times in the mostly rural area are in the two- to four-minute range, far better than before using the app, say fire officials. This emergency app is offering a big SOS button, which app users press to set the process in motion. Once the button is pressed, the NowForce reporting center will alert responders who are in the area of an emergency, including police, local security officials and United Hatzalah emergency rescue workers. The app shows them a map of where the incident took place and provides turn-by-turn instructions to get to the site. It also provides forms, updates and anything else connected to the specific incident, and lets responders take photos, videos and audio recordings of the incident. However, it is said that 80% of calls to the police centers and lines are false, which causes resources to be wasted: human, money, time and resources. The app is available for Android and iOS smartphones.

SmartSafe is a new solution, proposed in Ecuador, which seeks to minimize the kidnapping rescue time, using ICTs and the community as allies through their smartphones; a real-time, responsive "Citizen Warning and Feedback" mechanism in order to empirically reduce domestic crime, vandalism and kidnapping. This research focuses on designing and developing a distributed system that in real time notifies the nearby community (security agents, public transportation, stores, etc.) that a person has been kidnapped. All these are done through a new M2M protocol called Message Queue Telemetry Transport (MQTT) [MQT 15], ideal for small sensors where energy consumption is an important factor. It is a simple and lightweight publish/subscribe messaging protocol, designed for constrained devices and low-bandwidth, high-latency or unreliable networks. The design principles are to minimize network bandwidth and device resource requirements, while attempting to ensure reliability and some degree of assurance of delivery. These principles also turn out to make the protocol ideal of the emerging "machine-to-machine" (M2M) or "Internet of things" world of connected devices, and for mobile applications where bandwidth and battery power are at a premium [MQT 15]. In conclusion, MQTT is suitable to work with smartphones, whose processing power, sensors and communication types will complement the system requirements. For this, the development of an opportunistic application mobile will be a necessity and take into consideration the event known as "crowd sensing" [GAN 11]; it possibly ensures that the more the smartphones (help people), the greater the efficiency achieved by the system.

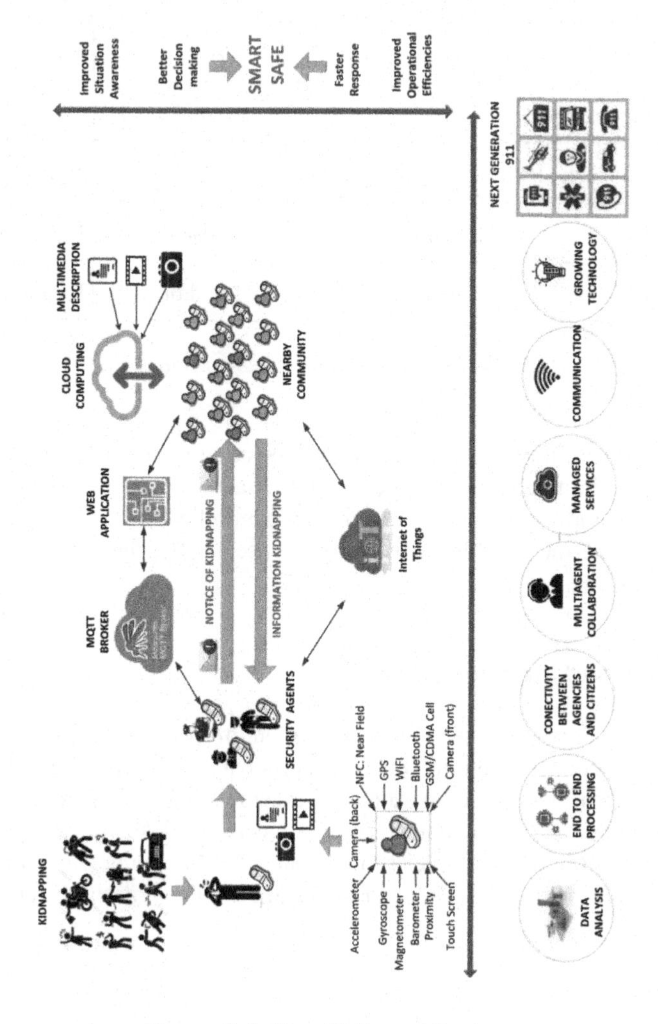

Figure 3.4. *SmartSafe architecture*

These proposals, both "I'm getting kidnapped" and NowForce, work with smartphones in a social setting but do not include the community as part of the solution. The goal is to merge new technologies, such as the Internet of Things (IoT [THE 06]), Cloud Computing (CC) [IBM 16] and smart cities [SMA 16], in order to rescue kidnapped people. Table 3.5 provides the main advantages and disadvantages of each solution.

Characteristics Project	I'm getting kidnapped	NowForce	SmartSafe
Smartphones as sensors	Yes	Yes	Yes
EWS	Yes	Yes	Yes
Community support	No	No	Yes
High rate of false alarms	Yes	Yes	Medium
Used time	Years	Years	Proposed
Operative System	Android	Android iOS	Heterogeneity

Table 3.5. *Comparison of safe projects*

Certainly, the ability to work under any operating system is a great advantage, since this lies in three points, one as a consequence of the other: increased community support, and therefore, a greater possibility to confirm a kidnapping, making lower the rate of false alarms, and consequently, the effectiveness of the entire system is higher.

3.4.3. *Roads and transportation*

Achieving smart cities with the integration of ICTs can make the existing infrastructure become the solution to current problems as transportation. Long delays to mobilize from one place to another, excessive fuel consumption, increased greenhouse gasses and even sometimes great economic losses make the intelligent transport systems (ITS) play a crucial role in our society, becoming a new requirement in orderly and automated cities. For example, the transportation of goods between cities, where the goal is to reduce congestion and CO2 emissions, and furthermore, improving security, which is currently a critical problem in several Latin American countries, without mentioning the stress it generates in people undergoing this problem daily. The most relevant projects in the field of transportation based on smart city technology are presented below:

– *Evaluating the feasibility of using smartphones for ITS safety applications [SER 13]*: the safety and comfort of driving can be improved with the use of intelligent transport; this work presents an application of security in a smartphone based on a protocol of dissemination of alert for calls eMDR. Using a smartphone, we can minimize the cost of hardware and remove certain barriers; users would have to install just one application on their smartphone, which is integrated with a navigation system allowing us to access road maps, current location and information about the route. This project has analyzed the behavior of the wireless channel and location (GPS) service under different conditions, showing the result of Car2Car communication, smartphones being able to obtain an optimal degree of connectivity and being quite accurate in certain types of driving safety applications.

– *Unlocking the smartphone's senses for smart city parking [JEA 16]*: to avoid spending more time in search of parking places that contribute to 30% of the total traffic. As is the case in New York, systems have been developed that provide information to the drivers about parking space availability. That is why this project is a system based on smartphone, relying on the sensors of the smartphone, the presence of WiFi and the cellular infrastructure, called SmartPark. SmartPark allows us to automatically detect when a user leaves a parking spot, indicating that it is available again. This application uses a combination of statistical analysis on the reading of sensors and a novel Random Forest-based classification algorithm. The experimental results allowed us to correctly detect the free parking 100% of the time and caused the cell phone battery level drop by only 4%.

– *Evaluation of smartphone performance for real-time traffic prediction [RAU 14]*: owing to improved technology of the sensor, navigation performance and connectivity of the network, the smartphone plays an important role in systems' advanced travel information (ATIS). In this study, the smartphone is actively involved in the real-time prediction of traffic networks compressed; it is in this way that ATIS becomes the key component for the efficient utilization of the infrastructure of the existing traffic, allowing the dissemination of information in real time to users in the form of maps of the network. The information provided is often delivered via smartphone, GPS devices or desktop applications. It allowed the users to choose less congested routes, and finally, decrease their travel time, improve comfort and reduce pollution and noise in congested sites.

– *E-roads*: this is a project being developed in Ecuador, which is based on improvised measures to analyze and optimize the incoming and outgoing heavy traffic in large cities. The vehicle traffic in Ecuador is a current problem that is getting worse, increasingly intense and chaotic, resulting in the anxiety and tension of drivers. Furthermore, the amount of misspent fuel increases the emissions of greenhouse gases, and causes excessive delays in mobilization and insecurity, which leads to economic losses. Through the definition of traffic design, based on a smart city, we will solve the existing problems not only in Ecuador, but anywhere in the world. Obtaining real-time traffic followed by immediate notifications to end users is part of this solution, as shown in the following figure.

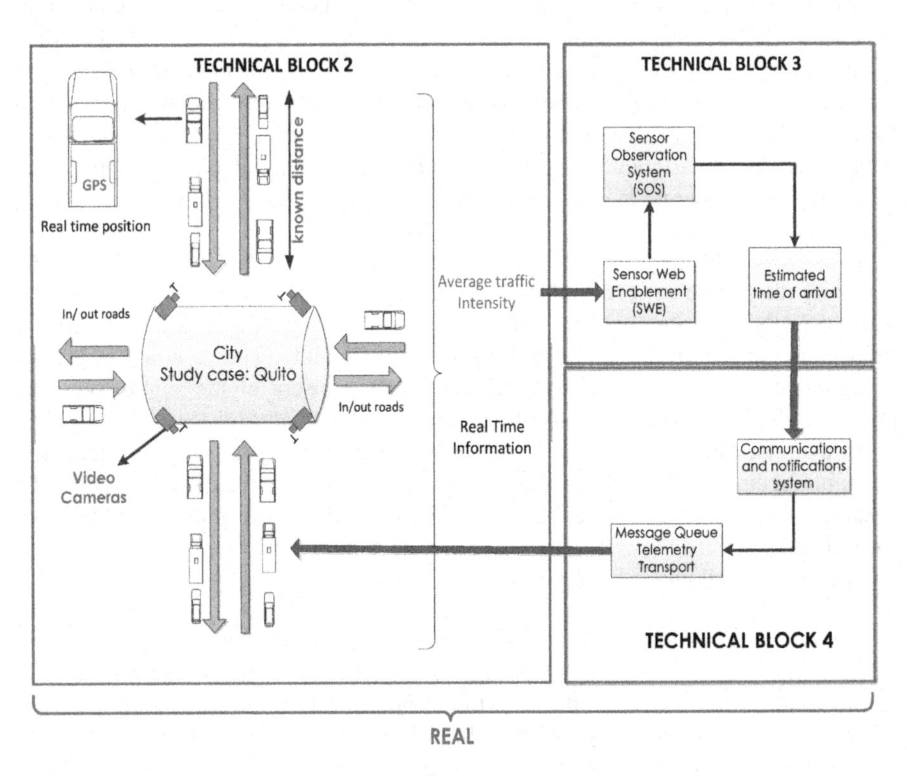

Figure 3.5. *E-roads architecture*

A comparison between the projects listed above is given in Table 3.6:

Project	Characteristics	Advantage	Disadvantage
Evaluating the feasibility of using smartphones for ITS safety applications	Use smartphone based on Android Uses smartphone based on Android Uses road maps Current location Path information Driving safety	ITS safety applications Channel Wireless Communications car to car (C2C)	To have a free map data source to avoid royalty issues 2.4 GHz band makes them not suitable for highly delay-constrained applications
Unlocking the smartphone's senses for smart city parking	Use smartphone Smartphone sensors Classification Algorithm WiFi, cellular infrastructure Minimal impact on battery life	Smart city parking Automatic transportation mode detection Minimum battery life impact	All crowdsourcing-based solutions, it suffers from a problem: the quality of the solution hinges on user adaptation while the user adaptation hinges on the quality of the solution
Evaluation of smartphone performance for real-time traffic prediction	Use smartphone, GPS devices, maps network Use Advanced Information System (ATIS) Forecast in real time	Real-time traffic prediction Decentralized infrastructure Traffic information, traffic conditions	Excellent performance smartphone Do not have a dynamic routing that is based on multi-horizon compressed prediction
E-roads	CEWS using smartphones Use smartphone, GPS devices, maps network Mobile application Forecast in real time	Real-time traffic prediction Heterogeneity using SWE Traffic information, traffic conditions Usable in other cities	Mandatory requirement Internet connection sensors The accuracy depends on the number of sensors

Table 3.6. *Comparison of roads and traffic projects*

3.5. Conclusions

Nowadays, cities around the world are designing new and innovative plans to transform into smart cities. That is to say, safer cities with less traffic, prepared for seismic events, and even more areas. The aim of this is to increase the quality of life of people, taking advantage of new ICTs and agile devices.

CEWS can potentially reduce risks and damage to properly handle four elements: risk knowledge, monitoring, response capability and warning communication. In each of these, science and technology can intervene to increase their profits. Today, there are new sensors: more flexible, faster, with better hardware capabilities, and with different communication capabilities. Furthermore, new protocols that enable real-time communications provide information on security, privacy, etc.

CEWS that use smartphones in their monitoring stage have access to a large number of sensors covering a wide area. Using methods of crowd sensing allows us to increase the reliability (reducing percentage error) of unreliable sensors with more sensors existing in the network. Thus, by applying three types of solutions in different fields (seismology, security and transport), it has been demonstrated how a CEWS delivers great benefits; in addition to the large use of smartphones in the current scenario, where it is possible to reuse components such as sensors at zero cost, it is also possible to get a lot of measuring and alert devices scattered throughout a whole territory in an economic and reliable way. Without a doubt, through this fusion of research, technology, and society, it is possible to achieve a better quality of life for the community.

To be able to make a city smart is important for the use of information and communication technologies allowing accessing, processing and delivering information in accordance with the urban environment that can later be used to solve problems. Obtaining this information requires a structured system that facilitates the effective data collection by creating architectures capable of interacting with the Internet of things, incorporating several devices, sensors and actuators in the urban context, in such a way that several architectures listed below have been created.

As we have seen through this chapter, there is no single way to solve a given problem (energy, transport, society, etc.). Depending on the scope of the problem, there are more and more proposals with different analyses, architectures, advantages and disadvantages. Therefore, the first stage of risk awareness within the Community Early Warning System is mandatory; collect and analyze the information, obtaining the unique requirements of each situation, and therefore, choose the best proposal to generate the best benefits. The biggest challenge that may be faced by the Community Early Warning System is to produce a solution a solution that allows anticipating critical events and, in addition, that uses reliable information in real time.

3.6. Bibliography

[ANS 14] ANSAR R., SARAMPAKHUL P., GHOSH S. *et al.*, "Evaluation of smart-phone performance for real-time traffic prediction", *17th International IEEE Conference on Intelligent Transportation Systems (ITSC)*, pp. 3010–3015, 2014.

[ASK 16] ASKSOURCE, "Access to emergency alerts for people with disabilities: recommendations for accessible emergency notification", available at: http://www.asksource.info/resources/access-emergency-alerts-people-disabilities-recommendations-accessible-emergency (accessed on 26 October 2016), 2016.

[BBC 16] BBC MUNDO, "María Elena Navas BBC: El drama de los niños desaparecidos de América Latina", available at: http://www.bbc.com/mundo/noticias/2013/11/131106_ninios_perdidos_desaparecidos_explotados_america_latina_men (accessed on 2 March 2016), 2016.

[BIL 16] BILICH K.A., "Child abduction facts", available at: http://www.parents.com/kids/safety/stranger-safety/child-abduction-facts/ (accessed on 3 March 2016), 2016.

[BRA 95] BRAUN T., DIOT C., HOGLANDER A. *et al.*, *An Experimental User Level of Implementation*, INRIA, 1995.

[CHA 09] CHANDY R., RITA A., SKJELLU Q., "QuakeCast: distributed seismic early warning", Caltech Undergraduate Research Journal, available at: http://www.rishichandy.com/files/QuakeCast_rishi_v3.pdf (Accessed on 15 December 2016), 2009.

[CIS 15] CISCO, Cisco visual networking index: global mobile data traffic forecast update, 2015–2020, White Paper, available at: http://www.cisco.com/c/en/us/solutions/collateral/service-provider/visual-networking-index-vni/mobile-white-paper-c11-520862.html (accessed on 26 October 2016), 2015.

[COU 11] COULOURIS G., DOLLIMORE J., KINDBERG T. *et al.*, *Distributed Systems. Concepts and Design*, 5th ed., Addison Wesley, 2011.

[COW 14] COWAN Y., O'BRIEN E., RAKOTOMALALA-RAKOTONDRANDRIA N., "Community-based Early Warning Systems: Key Practices for DRR Implementers", available at: www.fao.org/3/a-i3774e.pedf, pp. 51–53, 2014.

[DEL 01] DELIN K.A., JACKSONHDELIN K., JACKSON S., "Sensor web: a new instrument concept", *Symposium on Integrated Optics*, pp. 1–9, 2001.

[ELL 12] ELLIOT J., DAVIDSON J., Guidelines on Early Warning Systems and Application of Nowcasting and Warning Operations, Report, World Meteorological Organization, 2012.

[ERV 11] ERVASTI M., DASHTI S., REILLY J. *et al.*, "IShake: mobile phones as seismic sensors – user study findings", *Proceedings of the 10th International Conference on Mobile and Ubiquitous Multimedia*, no. 11, pp. 43–52, 2011.

[EST 15] ESTEVE M., Sistemas y protocolos de tiempo real, Polytechnic University of Valencia, 2015.

[EWC 06] EWC III, "Third International Conference on Early Warning (EWC III)", available at: http://www.unisdr.org/files/608_10340.pdf (accessed on 15 December 2016), pp. 1–10, 2006.

[FAU 11] FAULKNER M., OLSON M., CHANDY R. *et al.*, "The next big one: Detecting earthquakes and other rare events from community-based sensors", *ACM/IEEE International Conference on Information Processing in Sensor Networks IPSN'11*, pp. 13–24, 2011.

[FAZ 12] FAZIO M., PAONE M., PULIAFITO A. *et al.*, "Heterogeneous sensors become homogeneous things in smart cities", *Sixth International Conference on Innovative Mobile and Internet Services in Ubiquitous Computing (IMIS)*, pp. 775–780, 2012.

[FOR 16] FORBES, "Meet the free app that wants to keep you safe from kidnapping", available at: http://www.forbes.com/forbes/welcome/?toURL=http://www.forbes.com/sites/alexkonrad/2014/06/05/the-free-app-to-keep-you-safe-from-kidnapping (accessed on 3 November 2016), 2016.

[FUR 10] FURUKAWA, "Solution for Smart Cities", available at: http://www.pe.emb-japan.go.jp/jp/04%20_FISA_SmartCities_ES-Rev05s.pdf (accessed on 3 November 2016), 2010.

[GAN 11] GANTI R.K., FAN Y., HUI L., "Mobile crowdsensing: current state and future challenges", *IEEE Communications Magazine*, vol. 49, no. 11, pp. 32–39, 2011.

[GSM 16] GSMA, "Mobile Connectivity Index Launch Report", available at: http://www.gsma.com/mobilefordevelopment/programme/connected-society/mobile-connectivity-index-launch-report (accessed on 26 October 2016), 2016.

[HAN 12] HANCKE G.P., SILVA B.C., HANCKE G.P. Jr., "The Role of Advanced Sensing in Smart Cities", *Sensors*, vol. 13, no. 1, pp. 393–425, 2012.

[HER 11] HERNANDEZ-MUÑOZ J.M., VERCHER J.B., MUÑOZ L. *et al.*, "Smart cities at the forefront of the future internet", *The Future Internet*, Springer, Berlin Heidelberg, 2011.

[IBM 16] IBM, "What is cloud computing?", available at: https://www.ibm.com/cloud-computing/learn-more/what-is-cloud-computing/ (accessed on 27 October 2016), 2016.

[IGE 16] IGEPN, "Análisis del evento del 16 de Abril en base a la Experiencia de Grandes Terremotos y Tsunamis en Ecuador y Japón" available at: http://www.igepn.edu.ec/informacion-y-noticias/1368-analisis-del-evento-del-16-de-abril-del-2016-en-base-a-la-experiencia-de-los-grandes-terremotos-y-tsunamis-en-ecuador-y-japon, (accessed on 27 October 2016), 2016.

[INT 12] INTERNATIONAL FEDERATION OF RED CROSS AND RED CRESCENT SOCIETIES, "FRC: CEWS Guiding Principles", available at: http://www.ifrc.org/PageFiles/103323/1227800-IFRC-CEWS-Guiding-Principles-EN.pdf (Accessed on 3 November 2016), 2012.

[JAY 06] JAYASINGHE G., FAHMY F., GAJAWEERA N. *et al.*, "A GSM alarm device for disaster early warning", *First International Conference on Industrial and Information Systems*, pp. 383–387, 2006.

[JEA 16] JEAN-GABRIEL K., GENTIAN J., HADRIEN T. *et al.*, "Unlocking the Smartphone's Senses for Smart City Parking", *IEEE International Conference, In Communications (ICC)*, pp. 1–7, 2016.

[JOA 01] JOAQUÍN S.P., "Relojes lógicos", Departamento de Ingeniería de Sistemas Telemático, Universidad Politécnica de Madrid, 2001.

[JOH 10] JOHNSON J., MITCHELL H., LAFORCE S. *et al.*, "Mobile emergency alerting made accessible", *International Journal of Emergency Management*, vol. 7, no. 1, p. 88, 2010.

[JUN 15] JUNIOR D.D.S., CARDOSO I.F., MOMO M.R., "Tool-based mobile application applied to the monitoring system and flood alert", *Ninth International Conference on Complex, Intelligent, and Software Intensive Systems (CISIS)*, vol. 2015, pp. 348–351, 2015.

[MER 10] MERCY CORPS., "Practitioner`s Handbook Establishing Community Based Early Warning Systems", Mercy Corps and Practical Action, available at: http://repo.floodalliance.net/jspui/44111/1084 (Accessed on 15 December 2016), 2010.

[MIN 16] 20MINUTOS, "Cada año desaparecen más de 250.000 niños en Europa", available at: http://www.20minutos.es/noticia/2156144/0/desaparicion/ninos/europa/ (accessed on 3 March 2016), 2016.

[MQT 15] MQTT OFFICIAL SITE, "MQTT ORG: FAQ – Frequently Asked Questions | MQTT", available at: http://mqtt.org/faq (accessed on 20 January 2015), 2015.

[NOW 16] NOWFORCE, "Mobile Response Tools for Personnel & First Responders Situational awareness, everywhere", available at: http://www.nowforce.com/ (accessed on 27 October 2016), 2016.

[OPE 16] OPEN GEOSPATIAL CONSURTIUM, "Sensor Web Enablement", available at: http://www.opengeospatial.org/ogc/markets-technologies/swe (accessed on 27 October 2016), 2016.

[RAU 14] RAUF A., PARON S., SOHAM G. *et al.*, "Evaluation of Smart-phone Performance for Real-time Traffic Prediction", *17th International IEEE Conference on Intelligent Transportation Systems (ITSC)*, pp. 3010–3015, 2014.

[SER 13] SERGIO M.T., CARLOS T.C., JUAN-CARLOS C. *et al.*, "Evaluating the Feasibility of Using Smartphones for ITS Safety Applications", *IEEE 77th Conference In Vehicular Technology*, pp. 1–5, 2013.

[SMA 16] SMART CITIES, "Smart City: Vive la Ciudad Inteligente", available at: http://www.creatingsmartcities.es/ (accessed on 27 October 2016), 2016.

[TAN 08] TANENBAUM A., VAN STEEN M., *Sistemas Distribuidos: Principios y Paradigmas*, 2nd ed., Prentice Hall, 2008.

[THE 06] THE INTERNET OF THINGS, "IoT Council, a thinktank for the internet of things", available at: http://www.theinternetofthings.eu/ (accessed on 27 October 2016), 2006.

[TOR 13] TORNELL S.M., CALAFATE C.T., CANO J.C. *et al.*, "Evaluating the feasibility of using smartphones for ITS safety applications", *IEEE 77th Vehicular Technology Conference (VTC Spring)*, pp. 1–5, 2013.

[XIN 15] XINHUA F., YONGJIA Y., "Design and implementation for early warning system of airfield security defense", *IEEE ICICTA*, pp. 364–366, 2015.

[ZAM 14] ZAMBRANO A., PEREZ I., PALAU C. *et al.*, "Quake detection system using smartphone-based wireless sensor network for early warning", *IEEE International Conference on Pervasive Computing and Communications Workshops (PERCOM Workshops)*, pp. 297–302, 2014.

4

Generating Crisis Maps for Large-scale Disasters: Issues and Challenges

"We cannot stop natural disasters but we can arm ourselves with knowledge: so many lives wouldn't have to be lost if there was enough disaster preparedness."– Petra Nemcova, the supermodel who survived the 2004 Indian Ocean Tsunami.

Disaster is inevitable, and may strike the mankind any moment, anywhere. The after-effect of a disaster is destruction – of lives (human as well as non-human), properties

---

Chapter written by Partha Sarathi PAUL, Krishnandu HAZRA, Sujay SAHA, Subrata NANDI, Sandip CHAKRABORTY and Sajal DAS. This Publication is an outcome of the R&D work undertaken in the ITRA project of Media Lab Asia entitled. "Post-Disaster Situation Analysis and Resource Management Using Delay-Tolerant Peer-to-Peer Wireless Networks (DISARM)".

and infrastructures. To mitigate the post-disaster effects of a large-scale disaster is to ensure immediate response (rescue, relief and evacuation, if needed) at the right moment and at the right place. The above demand of the situation may be provided only if a channel for exchanging information between the affected community and the persons in charge of disaster risk mitigation and reduction could be established. This chapter describes novel strategies and their implementations that may be used for collecting on-site situational data in real time for preparing "localized" crisis maps that may carry more relevant information of the disaster-affected site. Incorporation of such a tool, as part of public safety network, may enhance the usefulness of such networks to cater the stakeholders better serving the affected community even after large-scale disasters.

4.1. Crisis mapping: "global" versus "local"

4.1.1. *Why crisis mapping?*

Crisis mapping is a relatively new concept in the literature of post-disaster rescue-relief operations and is introduced by the humanitarian organizations for aiding the post-disaster rescue-relief activities. For a few years now, crisis mapping services have been offered by many humanitarian organizations after large-scale disasters striking a region. Such services are now shaping the disaster response and relief operations for any kind of disaster that hits the mankind in any part of the universe.

DEFINITION 4.1.– Crisis mapping is a way of gathering, displaying and analyzing data in real time during a crisis, i.e. a natural disaster or social/political conflicts (violence, elections, etc.). Existing crisis mapping projects allows large numbers of people, including the public and crisis responders, to contribute information either remotely or from the site of the crisis. This kind of mapping increases situational awareness about the crisis of the disaster affected area.

In the case where a (natural) disaster strikes a region, the answers to the following questions are very critical for any disaster response authority in action:

1) Which of the roads are still passable?

2) Which patches of land can serve as helipads or drop zones?

3) Where are the landslides?

4) How many structurally sound buildings will aid workers find once they arrive?

Answering those questions is the goal of any crisis mapping, which nowadays has become an important assist to first responders[1] in any large-scale natural disasters, as observed during Nepal Earthquake in 2015[2].

1 A person whose job entails being the first on the scene of an emergency, such as firefighters or policemen or local volunteers.

2 http://www.bbc.com/news/world-asia-32603870

Such information may be better obtained through enhanced collaboration of government officials and agents with people at local level through the involvement of community-based organizations and NGOs. How a successful two-way communication between the affected people and the decision-makers could improve the rescue-relief operations can be observed in a study by *Internews* in collaboration with *School of International and Public Affairs (SIPA)* at Columbia University [REL 13]. We have also observed from a report from *UNISDR* that several means of communication method had been exercised before and during *Cyclone Phailin* in India (2013) including email, fax, phone as well as mobile phones sending SMS alerts to people, which helped the concerned country to reduce the life and property loss to a bare minimum during said disaster. The United Nations have considered this exemplary for disaster risk reduction and mitigation in the context of developing countries from South Asia as well as from the globe [UNE 13].

4.1.2. *Crisis mapping: working principle*

Crisis mapping applications mostly rely on crowd-sourced data from the stakeholders and the people nearby regarding the disaster situation analysis and damage assessment through online social networking services (Twitter/Facebook/...), news feed, SMS, etc., and represent the obtained information by overlaying the collected data in some form on a satellite or an online digital map like Google Earth or OpenStreetMap. Some well-known crisis map services include *Google Crisis Map*[3], *Google Crisis Response*[4], *Humanitarian OSM Crisis Mapping*[5] and so on. Such services often rely on online crowd-sourced data, and hence availability of the Internet connectivity in place is a mandatory requirement for these services being usable in the right place.

In Figure 4.1, we have hand-picked some of the sample crisis maps for a number of disasters of various nature (both natural and man-made) that had taken place in different parts of the globe. When investigating these crisis maps from various humanitarian agencies, we find that these maps embed a number of digital attributes in the disaster-affected region. The number and the nature of attributes vary from agency to agency and also on the type of disaster they would like to cater. Table 4.1 lists some of the attributes that are found over these maps.

3 https://google.org/crisismap/weather_and_events

4 https://www.google.org/crisisresponse/about/resources.html

5 http://wiki.openstreetmap.org/wiki/Humanitarian_OSM_Team

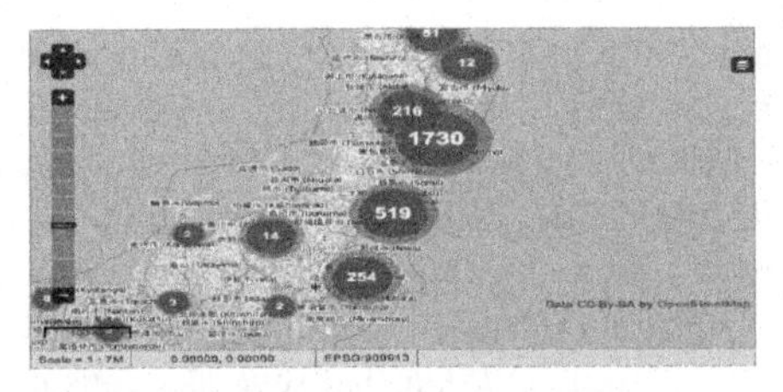

a) Japan Quake 2011 (Courtesy: [WWJ 11])

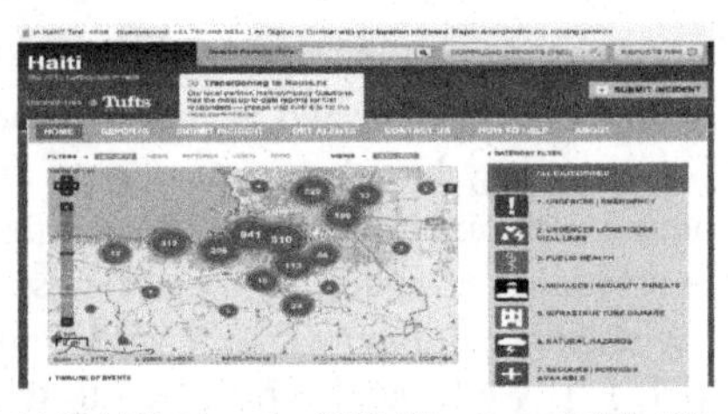

b) Haiti Earthquake 2010 (Courtesy: [MEL 10])

c) Uttarakhnad Flood 2013 (Courtesy: [MRU 13])

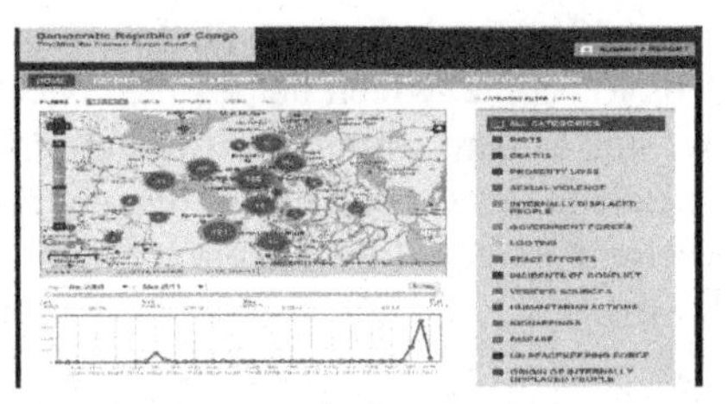

d) Eastern Congo Conflict (Courtesy: [HER 08])

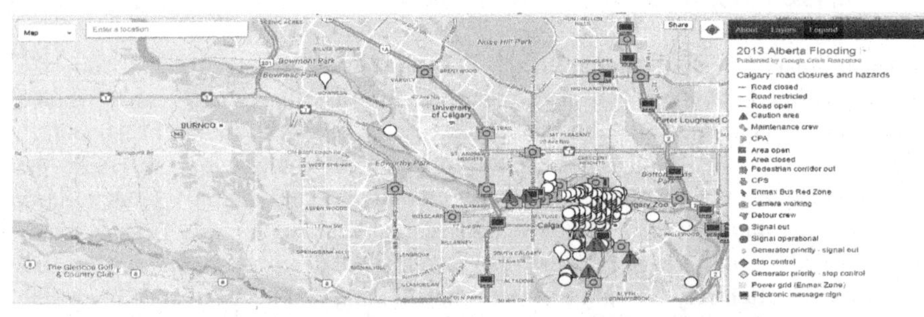

e) Alberta Flooding 2013 (Courtesy: [GOO 13])

f) California Wildfires 2012 (Courtesy: [GOO 12])

Figure 4.1. *Some example of crisis maps for disasters that occurred in different parts of the globe. For a color version of the figure, see www.iste.co.uk/camara/wireless3.zip*

Disaster	Earthquake in Haiti, 2010	Eastern Congo Conflict (2008–2011)	Alberta Flooding, 2013	Uttarakhand Flooding, 2013
Crisis Mapping Authority	Ushahidi	Ushahidi	Google Crisis Response	Google Crisis Response
Crisis Map Attributes	Emergency, Urgencies Logistics, public health, Security Threats, Infrastructure Damage, Natural Hazards, Resources Available	Riots, Deaths, Property Loss, Sexual Violence, Internally displaced people, Government forces, Looting, Peace Efforts, Incidents to conflicts, Verified sources, Humanitarian actions, Kidnappings, Disease, UN Peacekeeping force, Origin of internally displaced people	Road closed, Road Restricted, Road Open, Caution area, Maintenance crew, CPA, Area Open, Area Closed, Pedestrian corridor out, CPS, Enmax Bus Red zone, Camera Working, Detour Crew, Signal out, Signal operational, Generator priority - signal out, stop control, Generator priority - stop control, power grid (Enmax Zone), Electronic message sign	Rescue people list, Cleared rescue zones, People stranded, Medical Centers, Relief Camps, Shelters, Road Information, Mobile Access locations, Donation drop-off locations

Table 4.1. *Attributes for some well-known crisis maps.*

Figure 4.2 shows a schematic view of the workflow of existing principle of crisis mapping through crowdsourcing from online sources. In addition to this, additional information related to the demand and availability of resources of various nature at specific nodal points, the accumulation of victims trapped in various zones of the affected area may be marked on similar map-based interfaces for better decision support. But the major challenge behind feeding such a narrow level information in the system is to set up a direct information channel from the affected field to the information server. However, due to absence or disruption of a conventional network in the affected area, the flow of situational information from the field is either snapped completely or become inconsistent due to receiving partial and/or out of synchronization view from various zones of the affected area. In the following section, we would try to find out the post-disaster communication restoration scenario for last few large-scale disasters from different parts of the world that were very much in the news during the last decade, and also the means of alternative communication media that have been deployed in the interim.

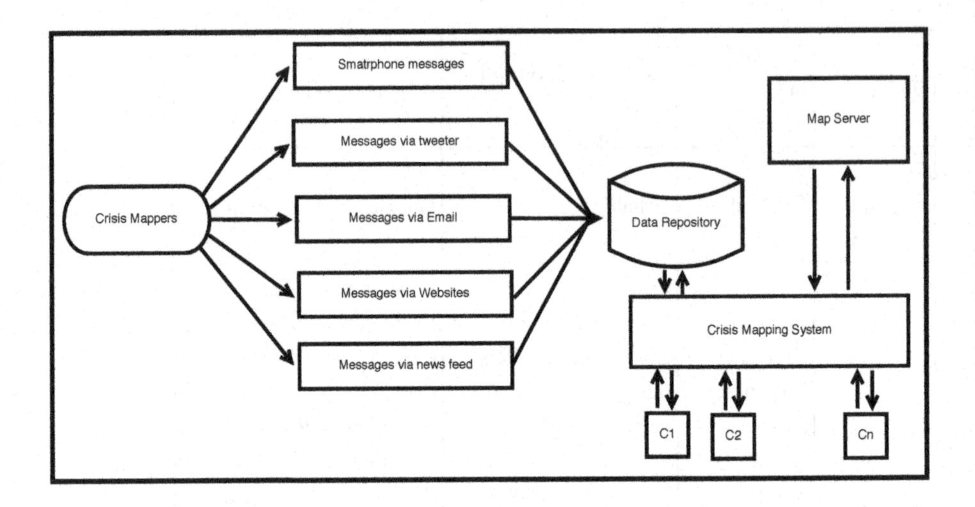

Figure 4.2. *A flow diagram illustrating the crisis mapping through online crowdsourcing*

4.2. Post-disaster communication revisited

4.2.1. *Post-disaster communication: state-of-the-art*

The user feedback during our field visits suggests that a lack of prompt information exchange during and after a large disaster makes the rescue-relief operations and evacuation operations ad-hoc, which often increases both the suffering of victims and rising death toll, even making the volunteers susceptible to become the victim of the situation. To ensure smooth dissemination of aforementioned crisis information even after a large-scale disaster, an integrated emergency information management system for collecting, processing and disseminating *in situ* information is highly warranted for "enhancing disaster preparedness for effective response and to Build Back Better in recovery, rehabilitation and reconstruction" [UNI 15b]. The desired system should gather information through participatory process; should be tailored for the needs of users at varying level, including social and cultural requirements; and needs to ensure that it remains effective and operational during and after disasters in order to provide life-saving and essential services. Since, disruption of conventional communication is very common phenomena after a large-scale disaster (due to destruction, power cuts, etc.), the system needs to operate seamlessly in the presence as well as in absence of conventional communication systems. In a nutshell, our objective in the present chapter is to propose an emergency information management system for rescue planning and relief need analysis that is capable of functioning in a situation where the conventional communication infrastructure is absent or highly disruptive.

Disaster Event	Ayla	Sandy	Gorkha Earthquake	Fukushima Daiichi Nuclear Disaster
Disaster Type	Cyclone & Flood	Tropical Cyclone	Earthquake	Earthquake, Tsunami, Nuclear Disaster
Year	May, 2009	October, 2012	April, 2015	March, 2011
Country	Bangladesh, India, Myanmar	USA, Canada	Nepal	Japan
Affected Population	2.3 million [BOS 09]	60 million	5.6 million [UNF 16]	0.4 to 1.0 Million [WIK 11, VER 12]
Time to Restore Electricity	More than 24 hours [INT 09]	More than 7 days [WIK 16]	couple of weeks [MIT 16]	One week [BEN 11, KAZ 12]
Time to Restore GSM/CDMA	3-4 days	4-7 days in some area [REA 12, CLA 12]	Few days [MIT 16], 38% of 2G and 4% of 3G network sites were down after the earthquake [KHA 15], The number of base stations has come down by half after the earthquake [SHR 15]	7 days [UEH 13]
Alternative Network Used	Amateur Radio [YAM 09]	Not Known	A radar detector, known as Finder by NASA [BAL 15], 35 satellite phones,10 BGAN terminals along with 10 laptops for the BGANs, 25 solar chargers for satellite phones and solar powered batteries [KHA 15]	Portable satellite Equipment, IPSTAR, UAV

Table 4.2. *Critical information related to some large-scale disasters observed during decade last*

Communication blackout is often an inevitable phenomenon of a large-scale disaster due to multiple reasons – damage of communication backbone, power crisis, signal jamming, etc. The affected community fails to connect the outside world, suffers from the insufferable pain, and often perishes being unattended even if the savior was in proximity. Such a feeling of being marooned from the mainstream leads to impatience and anger among the affected community towards the responders, the authorities and the outside world as a whole, which eventually affect the act of disaster response leading to further outrage and mistrust among the stakeholders. Were a communication, real or virtual, among the stakeholders persists in a situation like the one mentioned above, the disaster mitigation would have been easier and more effective. Table 4.2 highlights the damage scenarios and the associated information related to criticality of the situation in terms of the restoration electricity and conventional network connectivities for some of the large-scale disasters that hit the mankind during the last decade. The observation supports our previous assumption related to the disruption/unavailability of conventional network connectivities after large-scale disasters.

As observed from Table 4.2, for developed countries, the aforementioned problems are often solved partially by supplying mobile base stations and power generators on emergency basis to the affected region to partly mitigate the above problems. A typical example of such a solution may be found in the context of earthquake and tsunami in Japan in 2011 and its rescue-relief works [ITU 11]. Such solutions are quite difficult to adopt for developing and least-developed countries due to multiple reasons: (a) the supply of such costly equipment is usually very much limited for such countries; (b) the heavy-duty equipment like high-end power-generators and satellite antennas are very difficult to move to the core of the affected regions, as these places are often situated in the remotest part of the country; (c) due to security threat for the concerned country, the use of sensitive equipment like satellite phones are given controlled access among the top government officials and rescue-relief forces, thus keeping the affected common people away from the communication access. As a consequence, only a tailor-made solution could address the previously mentioned issue in the context of developing and least-developed regions; and mobile phones may play a significant role in that.

As found in a post-flood survey in Cambodia after the 2011 flood, the main medium of disaster-related information propagation during and after the flood was television and radio [ACT 12]. However, it was rightly pointed out that ownership of televisions is mostly restricted to wealthier members of the society, both the above mediums could hardly provide locally relevant information for the victims, such as location and time of relief distribution, closest medical help and so on. This is why the poorer community depends mostly on word-of-mouth via neighbors, relatives and other community members for receiving such information. However, the survey pinpointed that a larger fraction of the community indeed owns mobile phones; this medium of communication grossly underutilized during the said disaster. Experience from our field visits also suggests that mobile phones are broadly ignored as a medium of post-disaster communication in a larger context. The main reason behind the same is two-fold: (a) after a large-scale disaster, cellular connectivity becomes very much intermittent, and sometime completely absent even for days; (b) after a large-scale disaster, power crisis becomes a major concern for the affected community and the related stakeholders, as mobile phones are hardly usable even if the telecommunication network sustains. Similar observations can be found from reports regarding Uttarakhand Disaster 2013 in India [SAT 15] or Nepal Earthquake 2015 [UNI 15a] in a subcontinent context, and from technical reports from ITU-T in a global context [ITU 13]. As a consequence, the community and the associated stakeholders hardly consider mobile phones seriously for a post-disaster communication, although this medium has the capability of providing two-way communication between those affected and the emergency services, which the other modes of post-disaster communication mediums severely lack.

At present, countries in south-east Asia, alternative networks are created when specialized forces like army take the control of the situation. They use battery-powered satellite phones and associated transmitter-receiver devices carried in their backpack. The alternative solution is nice, except for the following criticism: (a) such networks remain fully under the control of the forces concerned, and are often out-of-reach of the other stakeholders; (b) forces often reach the site long time after the disaster, leading to complete communication blackout in the mean hours; (c) satellite phones have limited capacity and are often limited in numbers, and require the availability of satellites over the area during communication; (d) trained human resources are wasted for communication management, which could otherwise be employed for more fruitful purposes. Recently, amateur (HAM) radio service is being extensively used for the same. However, such operators are limited in number, and cannot cater every portion of the affected population. Secondly, requirement of strict licensing policies currently enforced by the authorities for the devices used by amateur radio operators restricts the possibility of extensive use of such devices in practice. Thirdly, the communication band allotted to amateur radio services is very narrow, and hence it is suitable only for voice-range communication. Digital transmission of multi-modal data over such a channel is not practical.

4.2.2. *Post-disaster communication: possible alternatives*

This is an era of smart devices. Our everyday life is now governed by smartphones and similar devices. Even decision-making is nowadays influenced by information obtained through smartphones. But such an essential support device for the modern era often becomes inoperative during and after large-scale disasters due to reasons mentioned earlier. The need of a robust public-centric communication for post-disaster situations has been requested by different humanitarian groups for a long time. In a survey paper in 2011, Franck Legendre has pointed out how opportunistic mode of communication may be used to uphold the public communication during large-scale disasters when normal communication infrastructure is destroyed, overloaded or non-existent in the first place [LEG 11].

A wireless communication system enables nodes to communicate while in motion. In an extreme case, when a communication infrastructure is unavailable, a mobile multi-hop ad-hoc network (MANET) is formed. Here each node functions not only as an end user but also as a router forwarding packets to and from other nodes to enable the multi-hop communication. However, in an ad-hoc network, increased node mobility causes frequent topology changes, link disruptions and even network partitions. In Figure 4.3(a), we can find a crisis situation, where mobile base stations are found either uprooted or in a non-functioning condition; however, there are enough number of mobile nodes in the region that is capable of creating a kind of mobile ad-hoc network in which an end-to-end path between any pair of nodes may be found at any point of time.

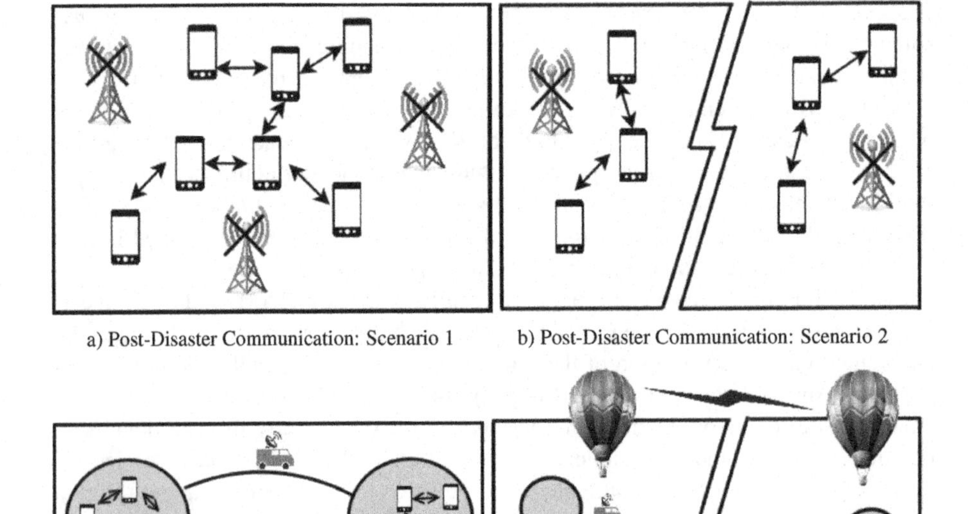

a) Post-Disaster Communication: Scenario 1

b) Post-Disaster Communication: Scenario 2

c) Post-Disaster Communication: Scenario 3

d) Post-Disaster Communication: Scenario 4

Figure 4.3. *Some post-disaster crisis scenarios and possible communication alternatives suggested against such scenarios*

In a sparse (i.e. with very low node density) MANET, these network partitions will become more prevalent. Hence a link between the nodes will frequently fluctuate due to node movement, obstruction to link by intervening bodies and shutdown of nodes to conserve power or to preserve secrecy. It results in intermittent connectivity and absence of an end-to-end path between the source and the destination at different times. In such scenario, conventional ad-hoc routing techniques fail to route messages to the destination since presence of an end-to-end route to the destination is one of their basic assumptions. Hence, messages get dropped when there is no existing route to destination. Hence, very low node density in MANETs may affect the delivery probability extensively, which in turn may affect the robustness of the network. In Figure 4.3(b), we may find a situation which is more similar to the previous one, except the fact that there are a pair of nodes for which no end-to-end ad-hoc connectivity is possible. The feature of intermittent connectivity in challenged scenarios triggers new challenges in packet routing. The solution to this problem has been dealt with considering the concept of a special kind of ad-hoc network known as Delay/Disruption-Tolerant Network (DTN).

For a decade, this concept has received increasing attention of both academia and industry. DTN is an attempt to extend the reach of traditional networking methods to situations where nodes are intermittently connected, and an end-to-end path from source to destination does not exist all the time. In such networks, instead of mostly fixed routes with almost fixed costs, routes with varying costs that build up over time exist owing to node movements. Messages need to be buffered at the nodes for long periods of time, waiting for a connection with their next relay node in the route to destination. In this way, these networks use store-and-forward mechanism to tolerate delivery delay of high orders. The amount of delay within acceptable limit may range from hours, days to even weeks depending on the application scenario such as extreme terrains, post-disaster communication and interplanetary, respectively. In DTN, end-to-end communication using TCP/IP protocol does not work due to long delay, asymmetric data rates and high error rates. Hence, the DTN architecture introduced a special layer called the bundle layer which is between the transport layer and the application layer of the TCP/IP protocol suite in order to provide end-to-end data transfer across heterogeneous networks. In DTN architecture, different transport protocols, as well as different network stacks, may be used in different network segments. DTNs support node-to-node retransmission of lost or corrupt data at both the transport layer and the bundle layer, ensuring node-to-node reliability.

However, if the geographical cover area to which the aforementioned DTN-based communication system has to cater is quite large, the delay that the data packets need to incur within the system may intolerably high, and the packet delivery ratio may as well become significantly poor. In addition, the participating DTN nodes may remain clustered at subregions with minimum node interaction across subregions. Figures 4.3(c) and 4.3(d) represent the scenarios of moderately large and extra-ordinarily large target cover areas, respectively. In both cases, mere DTN systems fail to provide satisfactory performance, and hence some hybrid DTN system (detailed shortly) may be tried as alternatives. As shown in the respective figures, in the former scenario, we may try using vehicles/UAVs with communication equipment as data carriers to augment the DTN system; and in the later scenario, some fast-deployable long-distance communication facilities like ballon/tower-mounted long-range WiFi/WiMax may be tried for further augmentation to the system. In the next section, we highlight some of the issues related to the design of large-scale DTN-based system to provide communication in a challenged scenario.

4.2.3. *Large-scale DTN systems for challenged scenarios*

Originally, DTN-based systems were targeted for enabling Interplanetary and Deep Sea Communication. In recent years, researchers have focused on the use of DTN in other challenged scenarios such as rural Internet and post-disaster communication. Such network systems use, in addition to DTN, infrastructure- based network components comprising of cellular network (available in isolated regions), satellite communication systems, mesh networks, data mules, etc. Any network in the challenged scenario uses hybrid network resources that can be organized into multiple tiers, such as DTN-enabled handheld devices at the bottommost tier. Some infrastructure-based components such as data mules (e.g. bus, ambulance, boat, UAV, etc.) as mechanical backhaul or wireless mesh networks at the middle tier and long-range WiFi/WiMax or satellite phones at the topmost tier. In the following, we briefly discuss about some of the existing large-scale DTN systems to understand their design limitations.

Systems based on mesh technologies are the Serval Project [GAR 11] (BatPhone provides a mesh mobile telephony platform) and RESCUE (a hybrid wireless mesh network for emergency situations) [DIL 08]. AirJaldi [SIN 12] and JaldiMAC [BEN 10] use long distance outdoor WiFi mesh to provide point-to- multipoint connectivity to broadband Internet in sparsely populated areas that suffer from large set-up delays. Although mesh technology is a good option for general challenged scenarios, for a large-scale disaster, the number of mesh devices increases drastically and ensuring line-of-sight features during their deployment is not always feasible. Project Daknet [PEN 04] and KioskNet [SET 06] use buses and cars as mechanical backhaul to ferry data to provide rural Internet but incur high latency. Systems Village Base Station (VBTS) [HEI 10] uses a low power GSM base station for rural telephony. Project LifeNet [MEH 11] and Twimight [HOS 12] provide connectivity under transient conditions using only handheld devices using DTN protocols. Use of vehicle-mounted communication devices as mechanical backhaul can be effective during disasters without any additional operating cost. The application of UAVs can overcome hurdles where physical connectivity is restricted due to landslides, etc. Contrary to the above approaches, Braunstein *et al.* [BRA 06] deployed a hybrid network architecture Extreme Networking System (ENS) for the support of a medical emergency response that consisted of three hierarchies: a WiFi network, a wireless mesh network and multiple backhaul networks, i.e. wired/ wireless/cellular/satellite). Except ENS and RESCUE, most of the systems mentioned above aims at enhancing rural infrastructure telephony, Internet, etc. Although wireless hybrid networking solutions have the potential to combine the advantages of different technologies to provide low-cost, scalable and reliable architectures. In Table 4.3, we have jotted down the above mentioned networking solution alternatives that may be found relevant ones in the present context, along with their underlying networking strategies and their limitation in connection with the present context.

System	Type of Networks	Description/Limitation
AirJaldi [SIN 12], JaldiMAC [BEN 10], RESCUE [DIL 08], Serval Project [GAR 11]	Long Distance WiFi Mesh	Provide Internet Connection covering a large area. The deployment overhead(cost & time) is high.
DakNet [PEN 04], KioskNet [SET 06]	DTN, GSM/GPRS	Provide rural internet connectivity using the concept of Data Mules. It incurs high latency.
LifeNet [MEH 11], Twimight [HOS 12]	DTN	Provide communication in transient environment. Delay becomes huge with use of only DTN nodes.

Table 4.3. *A comparative study of the DTN-based communication technologies being used for challenged situations.*

By noting all the solution technologies mentioned above and the criticism thereof, we would like to propose a multi-tier hybrid DTN solution that may better suit the situation mentioned in this context.

4.3. Proposed solution in a nutshell

In [SAH 15], we proposed a generalized architecture that enables a post-disaster communication network for catering a large coverage area. As a case study, we selected Sundarbans, West Bengal, India for a comparative analysis. On simulation-based comparison results with different cheaper wireless technologies and devices, we found that our proposed architecture would perform better than the other alternatives. In [PAU 15], we have presented a naive implementation of the architecture as a lab-scale testbed, performance of which establishes the proposed architecture as a decent alternative as challenged situation communication strategy.

4.3.1. *Multi-tier hybrid architecture for post-disaster communication*

Building an expert communication system with limited resources highly demands a proper planning about the usage of network resources. To explain the obviousness of a latency-aware hybrid network architecture for a challenged network, a comparative analysis has been made (Figure 4.4) with simulation results based on some simulation parameters as shown in Figure 4.5 of four restoration steps as shown in Figure 4.6 with their drawbacks.

	Scenario 1	Scenario 2	Scenario 3	Scenario 4
DTN	10 nodes / SP	10 nodes / SP	10 nodes / SP	10 nodes / SP
DB	×	1 unit / SP	1 unit / SP	1 unit / SP
DM	25 DMs Moving randomly along any existing pathway	25 DMs Moving randomly along any existing pathway	19 DMs for each SP moving DB to MCS	13 DMs for each SP excluding site SP of WT bounded from Group Center to MCS
WT	×	×		6 WTs - Randomly place these towers in such a way creating a connected network

Figure 4.4. *Utilization of various technologies for four different scenarios*

Simulation Specification			
Simulation Parameter	Value	Simulation Parameter	Value
Number of Shelter Points	19	Area of One Shelter Point	11 Sq Km
Packet Generation Load	10 Packets/Hr/DTN	TTL	200

Figure 4.5. *Simulation parameters for four different scenarios*

At the first step, no measurement has been taken after disaster as shown in Figure 4.6(a). Entire disaster hit area called Affected Area (AA) is divided into small zones with human belonging PDA devices, basically forms DTN with victims gathering at any public place; denoted as Shelter Points (SP) (it may be bus stoppages, gram panchayat office, school buildings, etc.) in our design. DTN-enabled PDA devices are moving randomly around the shelter point. In this case, some vehicles acting as Data Mule (DM) move among the SPs along the pathways in a random manner.

In the second step, one dropbox (DB) is placed at each SP to collect packets destined towards a fixed control center, i.e. Master Control Station (MCS) from DTN nodes and transmit them to data mules (DM) (Figure 4.6(b)). Here DTN-enabled smartphones are come in range of DB after certain interval. In this case also, the rescue vehicles (DM) are moving among the SPs along the pathways in random manner. In both the cases, we obtained lower delivery probability as shown in Figure 4.7.

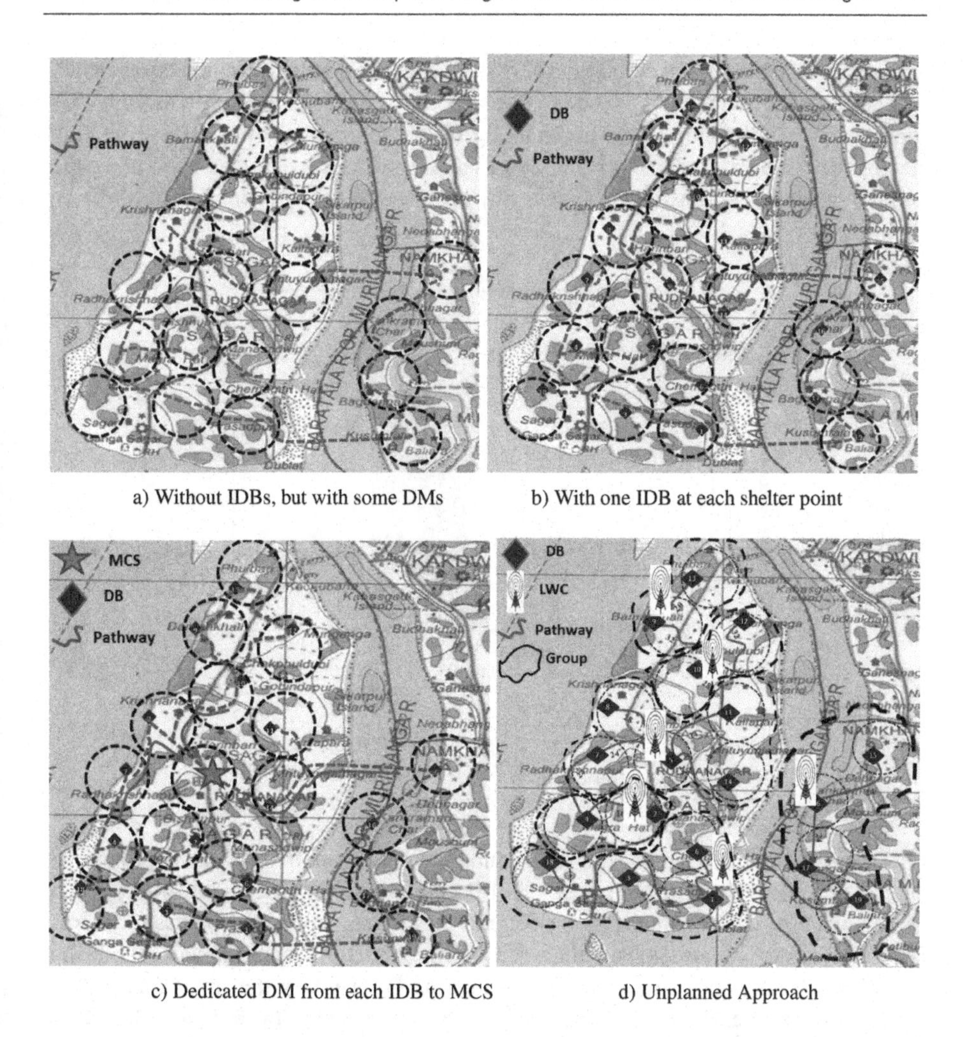

a) Without IDBs, but with some DMs

b) With one IDB at each shelter point

c) Dedicated DM from each IDB to MCS

d) Unplanned Approach

Figure 4.6. *Simulation under four different scenarios. For a color version of the figure, see www.iste.co.uk/camara/wireless3.zip*

In the third step, each SP has a dedicated DM with predefined unique trajectory (Figure 4.6(c)) which will move from DB to MCS. For this case, we are also getting poor performance as observed in Figure 4.7. To allow a dedicated data mule for each zone and assuming to have good pathways for all those devices to reach the MCS traversing such long distance are almost impractical after disaster.

The infeasible presumption and unsatisfactory performance of the third step compels us to set up a long-range infrastructure, i.e. some long-range communication units like long-range WiFi (LWC) which are placed initially by the brute-force method in the fourth step, creating a connected network for communication which was not possible practically by DMs. But selecting a best site to deploy these LWCs claims to be challenging while maintaining its line-of-sight feature. By simulating the aforesaid non-strategic approach, an unsatisfactory result enforces us to place the long-range LWCs with strategic manner by bringing forth our proposed architecture as shown in Figure 4.6(d).

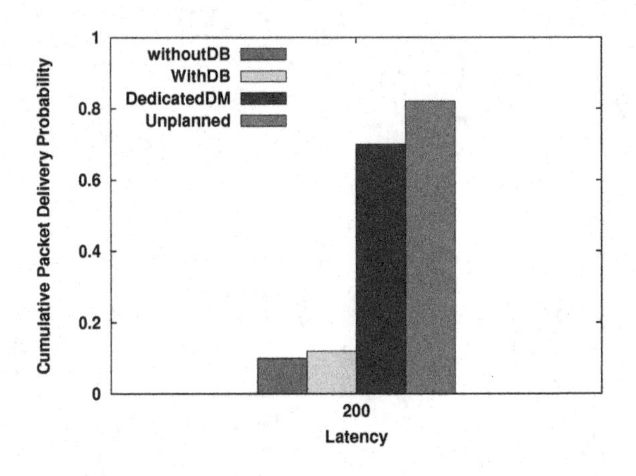

Figure 4.7. *Cumulative delivery probability for four different scenarios. For a color version of the figure, see www.iste.co.uk/camara/wireless3.zip*

The observation from the above simulation results motivated us to propose a multi-tiered hybrid solution in [SAH 15] using low-cost portable storage and communication devices that may be adopted to implement communication systems for post-disaster situations. In the proposed architecture, we claimed to use the following: (a) a pool of smartphones termed as DTN nodes that may function in opportunistic mode; (b) a pool of stationary devices termed as Information Drop Box (IDB) that is capable of storing and communicating information opportunistically; (c) a pool of vehicles (cars, boats, copters or UAVs) termed as Data Mules (DM) that are attached with specialized devices to store, carry and transfer information in an opportunistic mode; (d) a pool of portable and easily deployable (WiFi/WiMax) devices for a long-range communication termed as Long-range WiFi (LWC) that may be mounted at selected places. Figure 4.8 presents a schematic view of the hybrid opportunistic network architecture proposed in the aforementioned paper. Through extensive simulation of the proposed architecture using ONE simulator, we proved that the proposed architecture has the capacity of achieving nearly cent percent packet delivery rate within legitimate delay if some systematic approach is followed to deploy the corresponding system in the target area. We also proposed an empirical strategy of deployment of the same; however, there may be scope for improvement in the plan.

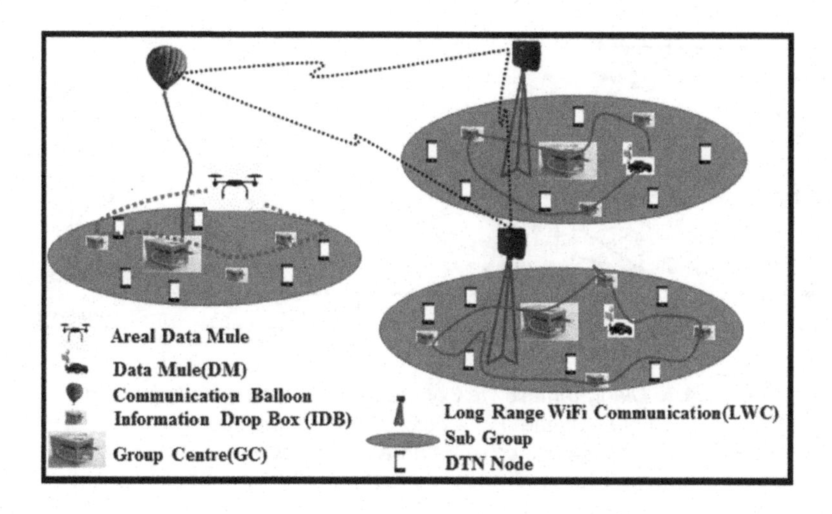

Figure 4.8. *A four-tier hybrid architecture for post-disaster communication. For a color version of the figure, see www.iste.co.uk/camara/wireless3.zip*

4.3.2. *Implementation & testbed*

In [PAU 15], we have presented a naive implementation of the architecture presented in formerly mentioned paper for implementing a lab-scale testbed for the delay-tolerant network research using low-cost devices. A schematic view of the said testbed implemented around our institute may be found in Figure 4.9.

In the said testbed, we used android-enabled smartphones as DTN nodes, laptops running Ubuntu Linux as IDBs, custom-built devices as DMs in which Raspberry Pi[6] running ARM Linux is used as computing unit, WiFi dongle is used for communication interface and power bank is used for power source. We may refer to Figure 4.10 for finding visuals for the custom-built device. In addition to the above, we used a pair of long-range WiFi devices mounted on guyed mast as LWC units. For seamless data transfer across such a wide variety of devices of heterogeneous nature and platform, we used a peer-to-peer file syncing tool called BitTorrent Sync[7]. The results in the aforementioned paper show more-or-less good data transmission performance over the network, which proves that the corresponding implementation of the proposed strategy of [SAH 15] has good hope in real-life application.

6 www.raspberrypi.org

7 http://blog.bittorrent.com/2015/03/03/sync-2-0-skip-the-cloud-share-direct/

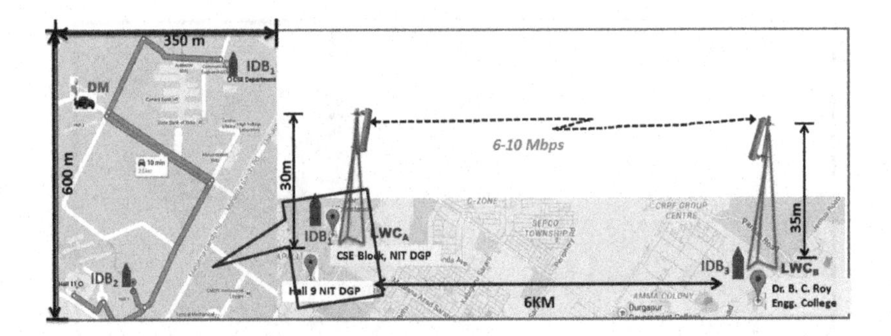

Figure 4.9. *Diagrammatic view of the PDM testbed installed in and around our institute. For a color version of the figure, see www.iste.co.uk/camara/wireless3.zip*

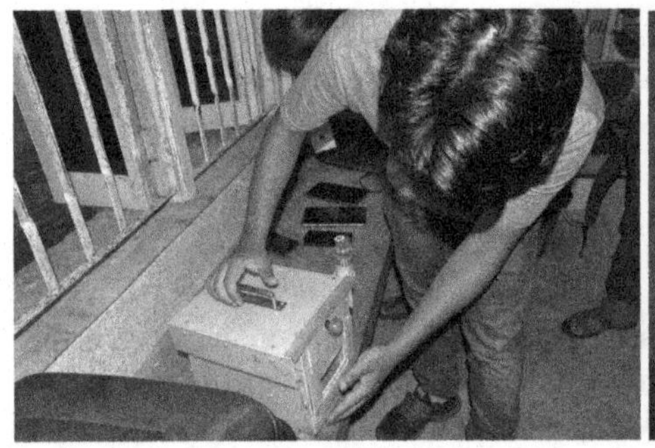

a) Custom-built Information Drop Box (IDB) unit

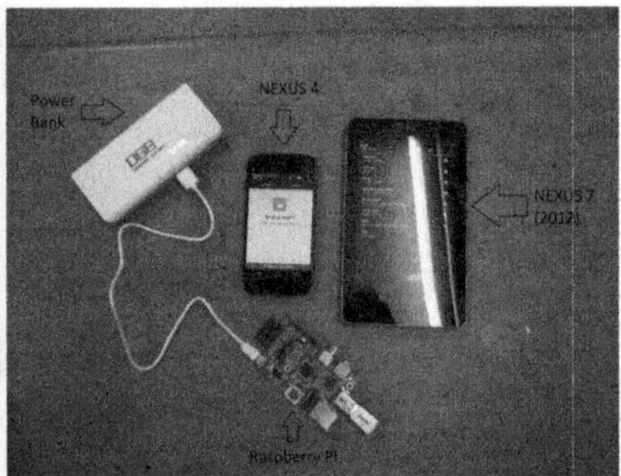

b) The components of IDB unit: A Raspberry Raspberry Pi, Power Bank, Micro-SD card and WiFi Dongle

Figure 4.10. *Device components used in our implementation of four-tier hybrid architecture for a post-disaster communication*

4.3.3. *Software suite*

A major challenge in developing the proposed system is to design and implement a software suite that would function seamlessly across the pool of devices mentioned above which are heterogeneous in genre and platform. Due to our assumption that the conventional communication system may be snapped during large-scale disasters, use of conventional TCP/IP protocol suite is not viable. We choose to use the DTN bundle protocol as the underlying networking technology. As protocol suite for the same, we may think of bundle protocol specification which is meant for delay-tolerant networks. Though bundle protocol gives a reliable way of communication between two devices, there are several issues that need to be addressed. Bundle protocol converts data into blocks of fixed size, called the "bundles". Here the question is, how to convert a file to bundles? In our case, it may be noted that files are of different sizes and will be transmitted from varying devices and platforms. Secondly, bundle protocol specification mentions nothing regarding the underlying communication framework and file transfer protocol above that. In our present study, we are about to present a protocol suite in a novel manner that would cater a wide class of devices and platforms communicating over a variety of wireless communication medium and provide some essential post-disaster management-related services seamlessly.

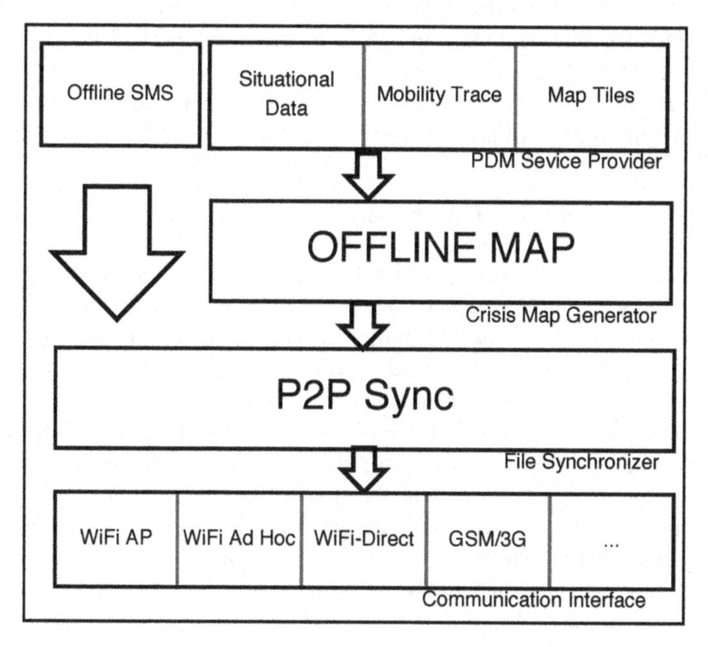

Figure 4.11. *Architecture for the proposed software system*

For developing the software suite, we use a protocol stack as shown in Figure 4.11, which will be implemented as part of the application layer of conventional TCP/IP stack. The bottommost layer of the proposed protocol stack deals with the communication interface through which the target devices would be connected. As viable communication alternatives, we tried multiple options. Most convenient way for establishing network communication among devices of heterogeneous nature and platform would have been WiFi ad-hoc. However, due to its power-hungry nature, it has been made obsolete by Google Inc. for android devices and will not be available near future [TRI 11] as a native application. However, the devices can be rooted/jail-broken to install WiFi ad-hoc interface into most of the standard smartphones through the use of CyanogenMod[8], an open-source operating system for smartphones and tablet PCs based on the Android mobile platform. The alternative solution provided for android devices, WiFi Direct, is not yet stable and suffers frequent crash and connection failures for heterogeneous devices[9]. The third option is to use the native WiFi AP mode available in the devices in an intelligent manner to connect themselves to each other, the detail mechanism adopted would be described in a later section. Finally, any other available communication interface like GSM/3G may also be used, if available.

On top of the underlying connection interface, a P2P sync application is used to synchronize user data across the nodes engaged in the connection. The P2P sync tool selected would take into account the users'/devices' role (authority level) in the system and the priority level of data files during process of file synchronization. The aforesaid *role-based file synchronization* enables the users to receive the files according to their role in the system in order of priority level of the files.

4.3.4. *Working principle of the proposed system*

Our objective in the present chapter is to propose a smart end-to-end information management system and associated strategies for information propagation and management to be used for rescue planning and relief need analysis. Such a system would ensure a persistent communication that may survive even after a large-scale disaster, would connect the affected community to the disaster managers and the outside world as a whole and would also take into account the issues related to partial and out of synchronization information from the field. Having set up such an information support system in action, the first responders synchronize their works with co-workers, and may plan the evacuation route for the victims from ground zero[10] and rescue-relief officials could efficiently plan the relief operations to best serve the victims to reduce their sufferings. Using our proposed system, the local and the state administration could collect crisis information from the stakeholders in order to pinpoint the rescue-relief need of the affected community. In a nutshell, our proposed system is designed to provide an alternative communication infrastructure with minimum possible investment towards the additional infrastructure that would survive even after a large-scale disaster, would ensure exchanging crisis-related information among the stakeholders at their maximum convenience, and as a by-product, would provide the stakeholders with a consistent view of the crisis situation and resource (human or material) demand and supply in an user convenient visual form maintaining their level of authority through mining and aggregating

---

8 http://en.wikipedia.org/wiki/CyanogenMod

9 https://code.google.com/p/android/issues/detail?id=25397

10 It is used in relation to earthquakes, epidemics and other disasters to mark the point of the most severe damage or destruction.

automatically the crisis information present in the system preserving the source anonymity. The proposed system comprises of a smart phone application, *Surakshit* (presently designed for Android platforms only, likely to be extended to other platforms in future) and a custom-built hardware suite, *XOB x.1*, precisely made for this purpose.

Using the following simple example scenario, we would like to demonstrate how our system would work. Modern smartphones could exchange media files with one another when they are in proximity, and many popular smartphone media file sharing applications (SHAREit, Xender, etc.) are developed based on that principle. Having installed Surakshit in his/her smartphone, a user could collect and exchange information through a convenient user interface. The messages will be forwarded automatically by the system to other users devices (users having Surakshit installed in their smartphones) whenever one comes in communication range of it, which in turn forward to some other devices until the message finds its final destination. The users of non-destination devices, which receive and retransmit the messages to the next devices, neither are interrupted nor could intercept the messages they receive. To strengthen and speed up the above mentioned store-carry-forward method of message propagation by the smartphones, XOB x.1 devices may be placed/preinstalled at strategic locations (Information Drop Box (IDB)), as well as associated with vehicles (cars/boats/copters/UAVs) which are then called Data Mules (DM). The resulting system may be visualized as one similar to a *Digital Postal System*. The IDBs collect and store the messages from the users that visit them like local post boxes. The DMs collect the messages from IDB and carry them to users like Mail Vans when they are far apart. The process may be explained pictorially through Figure 4.12.

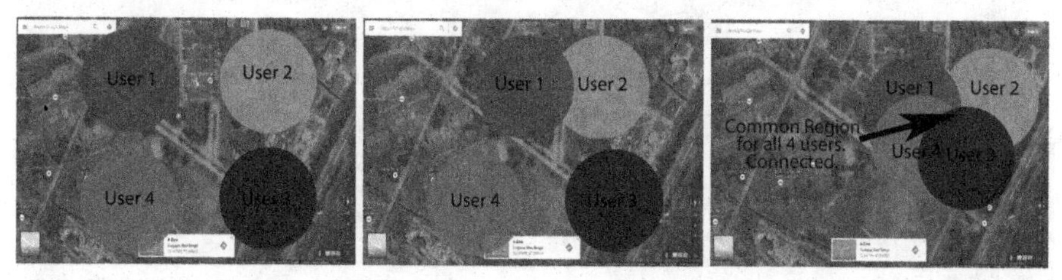

a) No interaction when nodes are away

b) Node interaction when a couple of nodes are in proximity

c) Node interaction when more than two nodes are in proximity

Figure 4.12. *An example scenario for user interactions during the trial. For a color version of the figure, see www.iste.co.uk/camara/wireless3.zip*

In Figure 4.12(a), the circles show approximate ranges of a cluster of users who are moving in an area represented by the underlying map view. Whenever a couple of users come in proximity, they form a small personal network and exchange messages seamlessly (refer Figure 4.12(b)). From Figure 4.12(c), it may be observed how all the four users in proximity of each other sharing a common region, and could form a larger personal network for sharing information.

To further illustrate the whole operation, we consider an example node interaction scenario as shown in Figure 4.13, where we have shown propagation of a message originated by user 1 and destined for user 6. Within the region shown in Figure 4.13(a), we may find five (05) users (having Surakshit in their respective smartphones) and a static XOB x.1. In Figure 4.13(b), we find user 1 meeting users 2 and 3, thus forwarding message to both of them. After a while, user 2 visits XOB x.1 and stores the message in the box. User 3 as well has the opportunity of meeting user 5, and thus forwards the message to user 5. Thus, as we observe in Figure 4.13(c), users 1, 2, 3 and 5 and the XOB x.1 have the replicas of the same message, and whoever has the opportunity of meeting with user 6 will deliver the message to it, which we find, as shown in Figure 4.13(d), to be XOB x.1 in the present scenario.

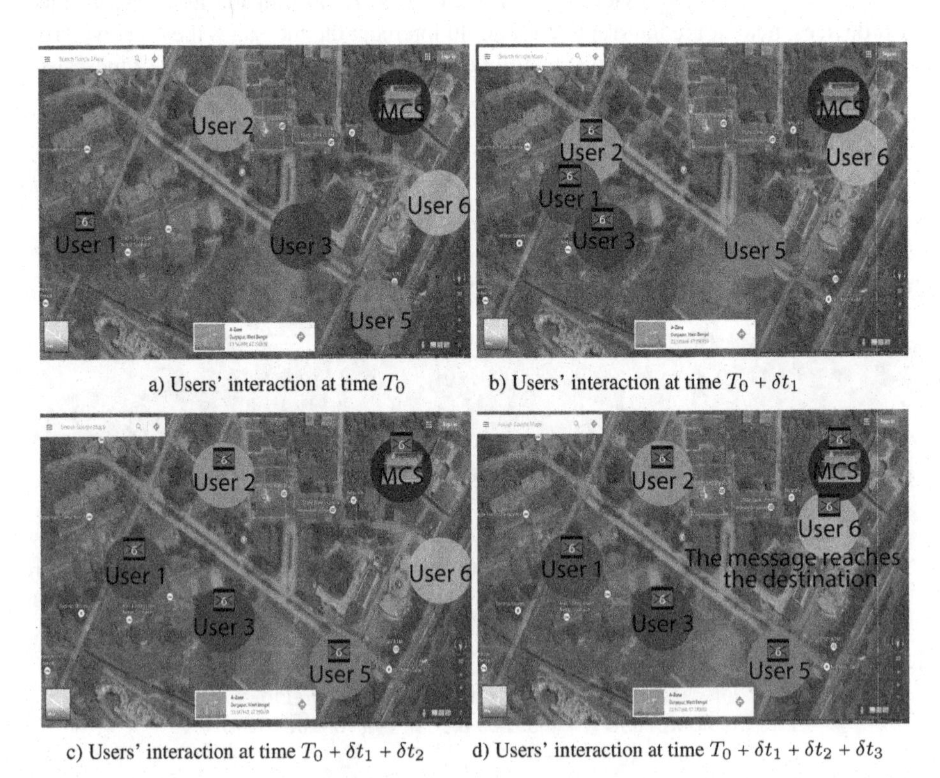

a) Users' interaction at time T_0 b) Users' interaction at time $T_0 + \delta t_1$

c) Users' interaction at time $T_0 + \delta t_1 + \delta t_2$ d) Users' interaction at time $T_0 + \delta t_1 + \delta t_2 + \delta t_3$

Figure 4.13. *An example scenario for user interactions for illustrating the working principle. For a color version of the figure, see www.iste.co.uk/camara/wireless3.zip*

In the way described above, the nodes exchange information among themselves. The software suite in the proposed system has the capability of exchanging information seamlessly without user intervention. User 1 in the previous example have created the message and instructed the system in his device to send the messages to user 6. The system determines the forwarding path intelligently, and eventually delivers the message to the device of user 6 when the first opportunity is found. More the contact opportunity among the users, faster the message delivery to the destination. Longer area coverage and faster delivery thereof may be achieved when IDBs are stationed at strategic locations where users often visit, and DMs are used to move messages across them.

4.4. Localized crisis mapping

4.4.1. *A system for offline crisis mapping*

When a disaster manager reaches the place of occurrence of a disaster, he has to report the following immediately after his arrival: (a) reporting at the place of occurrence (P.O.O.), (b) immediate requirement, (c) First Information Report (Calamity Assessment). For the above, often he has to wait for radio connectivity carried by the specialized forces (Disaster Response Force/Army). A step towards utilizing the smartphones of the disaster managers, ISRO has developed three smartphone applications to aid the officials in damage assessment. These applications have features like encrypted login, distress call (immediately after this is used) and emergency call (latitude, longitude, need medical help, need food and water). However, Internet connection is mandatory for all the above three applications to perform the basic functionality.

To bridge the gap between the reality and requirement, we propose our communication and service alternative that may aid the stakeholders even during a large-scale crisis situation. Using the proposed technology, the stakeholders will be able to create a network of their own, using even their basic android phones to create a virtual mesh through which they can communicate and transfer their messages hop-by-hop following a store-carry-forward paradigm.

With an aim in designing an end-to-end system for guiding the stakeholders in smart rescue planning and relief need analysis, we would like to implement an application for the mobile handheld devices (smartphones and tablet PCs). Our mobile application would help users in collecting, disseminating and visualizing the situational data in most convenient fashion for the users. However, due to intermittent and sparse connectivity in the affected area during the most critical hours, the data will be exchanged following a peer-to-peer mode of communication.

We choose three classes of data files that our mobile application will collect, store and exchange with peer nodes: (a) user mobility trails in the form of GPS log files to be used for evacuation/rescue route planning; (b) the situational (multi-modal) data in the form of user data files for rescue-relief need assessment; (c) map tiles for offline map visualization. User data files may be divided into multiple categories depending on the type of information they are holding (food requirement, medical urgency, reporting about victims, information related to shelter locations, etc.) in them. We implement the user data file exchange policy in our application following the strategy described in Algorithm 1. Every node evaluates the *Importance* value for all its data files based on their originating location and age, and determines a priority grading for all the files at its purview. When a node finds an opportunity to transfer user data files, it sends the files following the priority order thus obtained. For map tile sharing, the map tiles are ranked on the basis of situational data files generated from the subregion covered by the corresponding map tile, and the GPS dots obtained from the area given by the same. Higher ranked map tiles are sent prior to the lower ranked ones when the map tile transfer opportunity appears to a node.

For the sake of priority ranking of situational, let us define a measure *Importance* for the messages as defined below:

DEFINITION 4.2.– The *Importance* measure for a message is defined as the *Information* content holding by the message times an *Age Decay Factor* of the same. In other words, given the probability of occurrence of the message m_i is p_i and its creation time is t_i, then *Importance* for message m_i at the present moment t is given by the expression $-log(p_i) \times exp{-}(t - t_i)$.

Assigning *Importance* value to all the messages in a node, we ranked them following a predefined priority rule based on thus obtained importance value of the messages. The messages are exchanged following the priority order thus obtained. We observe from the definition above that importance of a message is proportional to the information content of the message given by $-log(p_i)$ in the present text, and hence the messages carrying higher information are supposed to get higher priority than the ones having lesser information content. On the other hand, the said measure decreases over time, and hence so is our priority rank of messages.

In [PAU 16], we discussed a crisis data dissemination strategy that would resolve a typical consistency-coverage trade-off described in the same. To do that, we assume that the area of interest is divided into subregions in the form of an $m \times n$ grid (refer components of Figure 4.14); N_{ij} denotes the number of messages of type T_i generated from the subregion R_j till time t from the beginning of time, j = 1 mn. Then, by time $T{+}\delta t$, the probability of occurrence of messages of type T_i from the subregion R_j may be taken as $p_{ij} = \frac{N_{ij}}{\sum_{i,j} N_{ij}}$. Following previous definition, the importance at time t of a message of type T_i from the subregion R_j generated at time t_0 may be expressed as $Imp(m_{ij}) = -log(p_{ij}) \times exp-(t - t_0)$. As observed in [PAU 16], ranking of messages through a priority order based on the *Importance* value defined here ensures to some extent the forwarding of newer messages ahead of older ones, and at the same time giving higher priority to messages originated in sub-zones from which the reporting is less frequent.

When data transfer opportunity appears to a node, the node send the data files in the order as follows: (a) mobility trace data (GPS log files), (b) situational information related data (user data files) and (c) map tiles. The empirical justification behind the strategy is the intuitive reasoning about the order of necessity for the corresponding information from user perspective: evacuation being the highest priority job during an emergency situation, followed by the relief service to the victims. For preparing smart evacuation planning, user mobility traces are required. For relief need analysis, situational messages are needed. Finally, for better presentation of both the above, map tiles need to be shared, if not already present in the users' device. A systematic step-by-step description of the data exchange strategy is given in Algorithm 1.

step 1: if *Mobility trace data files yet to be synced* **then**
  Sync mobility trace files.
end
step 2: if *Situational data files yet to be synced* **then**
  - Sort the situational data files according to the **Importance** (measure) value.
  - Sync the situational data files following the order of their priority thus obtained.
end
step 3: if *Map tiles yet to be synced* **then**
  - Sort the map tiles according to the number of GPS dots as well as the number of situational data files being geo-tagged with it.
  - Sync the map tiles following the priority order thus obtained.
end

Algorithm 1: Data sharing strategy followed by the nodes when they meet with each other during our experiment

For experimental validation of the crisis mapping strategy described above, the software system shown in Figure 4.11 is developed as a smartphone application suite. For exchanging user-collected data files, we felt that peer-to-peer file synchronization tool would be the best alternative. However, ready-made peer-to-peer sync tools would exchange data and other user files in the way they appear to the system (Epidemic Routing) without considering the "importance" of the file to the user. This is why, we implemented a custom-built peer-to-peer sync tool that takes care of the "importance" of the user files following Algorithm 1. For testing experimentally the functioning of the proposed strategy for localized crisis mapping, we used an experiment plan as illustrated in Figure 4.14.

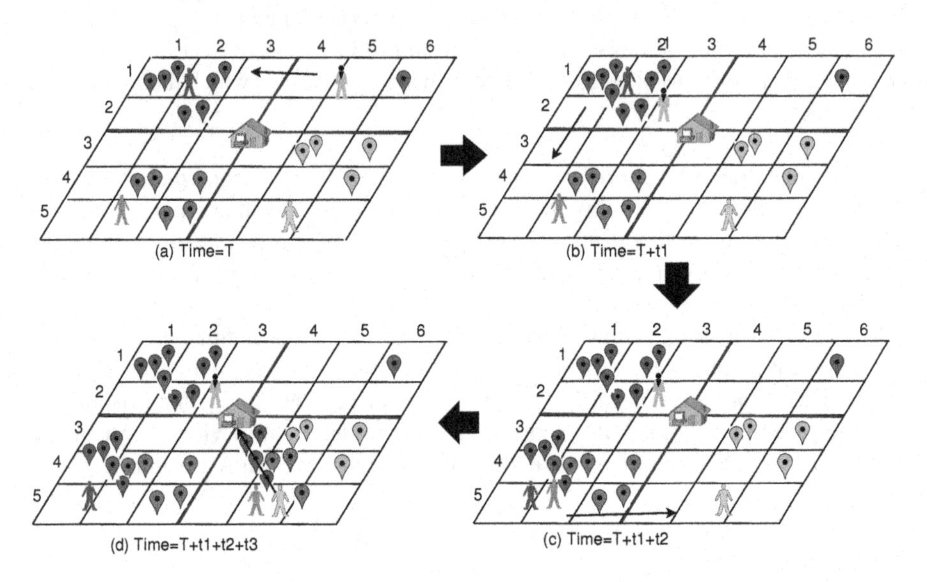

Figure 4.14. *Experiment plan using four dynamic nodes and one static node (IDB) for the purpose of demonstrating the development of the localized crisis map in the nodes through sharing of crisis information as they meet [Courtesy: [PAU 16]]. For a color version of the figure, see www.iste.co.uk/camara/wireless3.zip*

In Figure 4.15, we have presented an idealized version of a localized crisis map that we would like each node to sharing through offline exchange of local crisis data. Initially, at the beginning of experiment, four mobile nodes, referred node A (gray node) through node D (yellow node), have only their share of the information in their respective smartphones: (a) their own data files, (b) their own GPS trails and the map tiles of the neighborhood of their starting location. The static node, referred E placed near the hut in the middle, was completely blank at the beginning. The mobile nodes move in the way as shown in Figure 4.14, and wait during each node contact for a fixed time duration, when we expect the crisis data and other associated files are exchanged among peer nodes.

In Figure 4.15, we observed local crisis maps visible to all the five nodes described above after each node contacts. The result suggests that more important information contents are propagating to the next level ahead of the less important ones. Since we have given maximum emphasis on evacuation route information, user mobility traces for all the users are shared with highest priority. Among the situational data files, the medical and victim-related data has given higher importance. So, such data from all the nodes reached the static node. Since, map tiles are not given higher importance over the others, they are likely to reach the destination after the other files are transferred completely. Since, in our experiment scenario, the node contact durations were kept very small intentionally, very few map tiles reached to other nodes; no map tile reaches the static node. The final form of the local crisis map at each node matches with our expectation.

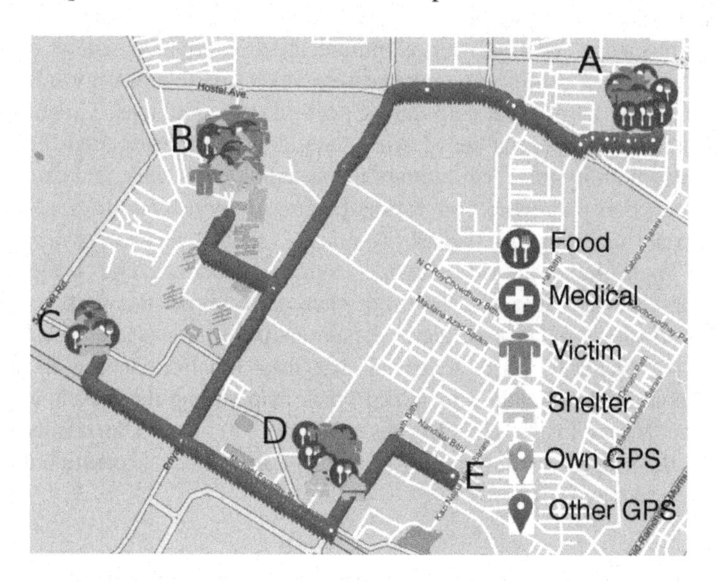

Figure 4.15. *An ideal version of the localized crisis map for a hypothetical affected region [Courtesy: [PAU 16]]*

4.5. Concluding remarks

We have observed through our discussion that crisis mapping is becoming an indispensable tool for shaping the disaster risk mitigation over the last decade. The present form of crisis mapping relies on collection of crisis data through social media, through SMSs from crisis mappers around the affected region, collecting crisis data from news feed and so on. All of the above form of data feeding requires the availability of a reporting agent who has the access to the infield crisis information, and at the same time has some means for sending that information to the online server in some form. In case of a large-scale disaster striking a large area, either or both the above may not be available immediately after the crisis, and the corresponding crisis mapping agencies have to rely on information obtained from the surrounding. This often would hamper the true objective of crisis mapping. For resolving the above issue, we need to find a mean for alternative communication strategy through which crisis mapping services could collect crisis data from the true stakeholders. We observed that special purpose forces such as army or disaster-response forces have their own ways of establishing some alternative form of communication, but ordinary victims and other stakeholders have very little access to those networks. Voluntary services like amateur radio nowadays provide great services towards a post-disaster communication establishment. However, amateur radio operators are very few in number, and they often reach the destination from outside, and hence would come into action after passing of valuable hours/days. Having established a post-disaster communication alternative solely relying on low-cost devices that common people possess (like smartphones nowadays) in its natural form would be of great help during post-disaster situations. Creating a crisis map solely through the exchange of crisis data along with associated files such as map tiles, user mobility trace, etc. among the stakeholders locally, and extending the same to global crisis map when conventional connectivity resumes may add a feather in cap of crisis mapping for disaster response. In this chapter, we proposed a DTN-based multi-tiered hybrid network using smartphones and some low-cost fast-deployable devices that may be used when conventional network connectivity snapped as an effect of an emergency situation like a large-scale disaster, and a framework for localized crisis mapping that may function on top of aforesaid network alternative to provide a localized crisis map among the stakeholders when they have no access to the global crisis mapping services. The results obtained suggest that the proposed solution may provide an alternative successful solution for the associated problems.

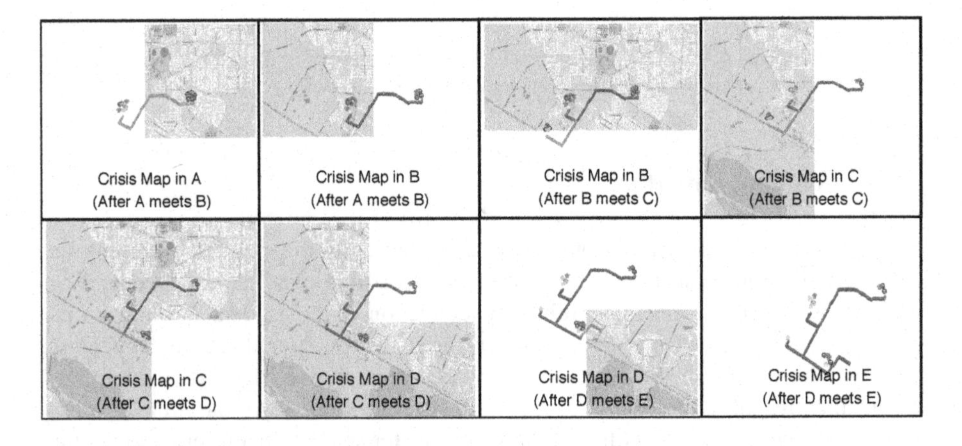

Figure 4.16. *Local crisis maps are visible to various nodes as they move following the plan as referred in Figure 4.14 [Courtesy: [PAU 16]]. For a color version of the figure, see www.iste.co.uk/camara/wireless3.zip*

4.6. Bibliography

[ACT 12] ACTION AID, ASIAN DEVELOPMENT BANK, "Cambodia - Post-Flood Relief and Recovery Survey", Survey Report, Action Aid, Asian Development Bank (ADB), DanChurchAid/ACT Alliance, Danish Red Cross, Save the Children, UNICEF and World Food Programme, available at: https://www.adb.org/sites/default/files/project-document/73688/46009-001-cam-dpp.pdf, 2012.

[BAL 15] BALLARD B., "Nepal earthquake: The technology helping to save lives", ITProPortal, available at: http://www.itproportal.com/2015/05/13/nepal-earthquake-the-technology-helping-to-save-lives/, 2015.

[BEN 10] BEN-DAVID Y., VALLENTIN M., FOWLER S. *et al.*, "JaldiMAC: taking the distance further", *Proceedings of the 4th ACM Workshop on Networked Systems for Developing Regions*, NSDR 2010, ACM, pp. 2:1–2:6, 2010.

[BEN 11] BENFIELD A., "Tohoku Earthquake & Tsunami Event Recap Report", available at: http://thoughtleadership.aonbenfield.com/documents/201108_ab_if_japan_eq_tsunami_event_recap.pdf, 2011.

[BOS 09] BOSTON.COM, "The Big Picture: Cyclone Aila", News Stories in Photograph, available at: http://archive.boston.com/bigpicture/2009/06/cyclone_aila.html, 3 2009.

[BRA 06] BRAUNSTEIN B., TRIMBLE T., MISHRA R. *et al.*, "Challenges in using distributed wireless mesh networks in emergency response", *3rd International ISCRAM Conference*, pp. 30–38, 2006.

[CLA 12] CLARK J., "Hurricane Sandy doubled failures in US internet infrastructure", ZDNET, available at: http://www.zdnet.com/article/hurricane-sandy-doubled-failures-in-us-internet-infrastructure/, 2012.

[DIL 08] DILMAGHANI R.B., RAO R.R., "A wireless mesh infrastructure deployment with application for emergency scenarios", *Proceedings of the 5th International ISCRAM Conference*, Washington DC, May 4–7 2008.

[GAR 11] GARDNER-STEPHEN P., The Serval Project: Practical Wireless Ad-hoc Mobile Telecommunications, Technical Report, Flinders University, Adelaide, South Australia, 2011.

[GOO 12] GOOGLE CRISIS MAP, US Wildfires: California, available at: http://google.org/crisismap/2012_us_wildfires, 2012.

[GOO 13] GOOGLE CRISIS MAPS, Albesta flooding map, available at: https://google.org/crisismap/2013-alberta-floods, 2013.

[HEI 10] HEIMERL K., BREWER E., "The village base station", *Proceedings of the 4th ACM Workshop on Networked Systems for Developing Regions*, ACM, p. 14, 2010.

[HER 08] HERSMAN E., "Ushahidi Deploys to the Congo (DRC)", available at: https://www.ushahidi.com/blog/2008/11/07/ushahidi-deploys-to-the-congo-drc, 2008.

[HOS 12] HOSSMANN T., SCHATZMANN D., CARTA P. *et al.*, "Twitter in disaster mode: smart probing for opportunistic peers", *Proceedings of the third ACM international workshop on Mobile Opportunistic Networks*, ACM, pp. 93–94, 2012.

[INT 09] INTERNATIONAL FLOOD NETWORK, "Impact of Cyclone AILA", Technical Report, International Flood Network, Infrastructure Development Institute-Japan, available at: http://www.internationalfloodnetwork.org/aila.htm, June 24 2009.

[ITU 11] ITU, "Japan after the earthquake and tsunami: update on the restoration of telecommunication and broadcasting services", available at: http://www.itu.int/net/itunews/issues/2011/06/32.aspx, July/August 2011.

[ITU 13] ITU, "Technical Report on Telecommunications and Disaster Mitigation", available at: http://www.itu.int/dms_pub/itu-t/opb/fg/T-FG-DRNRR-2013-PDF-E.pdf, June 2013.

[KAZ 12] KAZAMA M., NODA T., "Damage statistics (Summary of the 2011 off the Pacific Coast of Tohoku Earthquake damage)", 2012, Special Issue on Geotechnical Aspects of the 2011 off the Pacific Coast of Tohoku Earthquake.

[KHA 15] KHANAL A.R., "Nepal's Experience in Responding to a Disaster: A Telecommunication/ICT sector perspective", Technical Report, available at: https://www.itu.int/en/ITU-D/Regional-Presence/AsiaPacific/Documents/Events/2015/August-RDF2015/Session-4/S4_Ananda_Raj_Khanal.pptx, 2015.

[LEG 11] LEGENDRE F., "30 years of ad hoc networking research: what about humanitarian and disaster relief solutions? What are we still missing?", *Proceedings of the 1st International Conference on Wireless Technologies for Humanitarian Relief*, ACWR 2011, New York, ACM, p. 217, 2011.

[MEH 11] MEHENDALE H., PARANJPE A.,VEMPALA S., "LifeNet: a flexible ad hoc networking solution for transient environments", *Proceedings of the ACM SIGCOMM 2011 Conference*, SIGCOMM '11, New York, ACM, pp. 446–447, 2011.

[MEL 10] "Crisis Mapping Haiti: Some final reflections", available at: https://www.ushahidi.com/blog/2010/04/14/crisis-mapping-haiti-some-final-reflections, 2010.

[MIT 16] MITCHELL R., "Restoring the Internet in Nepal, one year after the quake", APNIC, available at: https://blog.apnic.net/2016/05/25/restoring-internet-nepal-one-year-quake, 2016.

[MRU 13] MRUNAL [Disaster] cloudburst, Himalayan tsunami in Uttarakhand, "Dopper weather", available at: http://mrunal.org/2013/09/disaster-cloudburst-himalayan-tsunami-uttrakhand-dopper-radar.html, 2013.

[PAU 15] PAUL P.S. *et al.*, "Challenges in designing testbed for evaluating delay-tolerant hybrid networks", *IEEE PerCom (WiP)*, pp. 280–283, 2015.

[PAU 16] PAUL P.S., DUTTA H.S., GHOSH B.C. *et al.*, "Offline crisis mapping by opportunistic dissemination of crisis data after large-scale disasters", *2nd ACM SIGSPATIAL International Workshop on Emergency Management using GIS (EM-GIS 2016)*, San Francisco, California, available at: http://www.dviz.cn/em-gis2016/files/EMGIS_2016_paper_17.pdf, 2016.

[PEN 04] PENTLAND A., FLETCHER R., HASSON A., "DakNet: rethinking connectivity in developing nations", Computer, vol. 37, no. 1, pp. 78–83, doi: 10.1109/MC.2004.1260729, 2004.

[REA 12] REARDON M., "Hurricane Sandy disrupts wireless and Internet services", CNET, available at: https://www.cnet.com/news/hurricane-sandy-disrupts-wireless-and-internet-services, 2012.

[REL 13] RELIEF WEB, "Communicating during Disasters: Examining the Relationship between Humanitarian Organizations and Local Media", Internews Network and Columbia University, available at: http://reliefweb.int/report/world/communicating-during-disasters-examining-relationship-between-humanitarian, 2013.

[SAH 15] SAHA S. *et al.*, "Designing delay constrained hybrid ad hoc network infrastructure for post-disaster communication", *Ad Hoc Networks*, vol. 25, Part B, pp. 406–429, 2015, New Research Challenges in Mobile, Opportunistic and Delay-Tolerant NetworksEnergy-Aware Data Centers: Architecture, Infrastructure, and Communication.

[SAT 15] SATENDRA, GUPTA A.K., NAIK V.K. *et al.*, "Uttarakhand Disaster 2013", National Institute of Disaster Management, Ministry of Home Affairs, GoI, available at: http://nidm.gov.in/books.asp, 2015.

[SET 06] SETH A., KROEKER D., ZAHARIA M. *et al.*, "Low-cost Communication for Rural Internet Kiosks Using Mechanical Backhaul", *Proceedings of the 12th Annual International Conference on Mobile Computing and Networking*, MobiCom '06, New York, ACM, pp. 334–345, 2006.

[SHR 15] SHRESTHA S., "The subscriber cannot be reached", News Report, Nepali Times, available at: http://nepalitimes.com/article/nation/connectivity-problem-post-quake, 2525, 2015.

[SIN 12] SINGH R.D., "Wireless for communities: empowering communities through wireless connectivity", *IEEE Internet Computing*, vol. 16, num. 3, 2012.

[TRI 11] TRIFUNOVIC S. *et al.*, "WiFi-Opp: Ad-hoc-less opportunistic networking", *Proceedings of the 6th ACM Workshop on Challenged Networks*, CHANTS '11, New York, ACM, pp. 37–42, 2011.

[UEH 13] UEHARA N., "Be Prepared!—Lessons learned from the Great East Japan Earthquake and tsunami disaster", *Japan Medical Association Journal*, vol. 56, no. 2, available at: https://www.med.or.jp/english/journal/pdf/2013_02/118_126.pdf, pp. 118–126, 2013

[UNE 13] UNEP, Cyclone Phailin in India: early warning and timely actions saved lives, Technical Report, India Environment Portal, available at: www.unep.org/pdf/UNEP_GEAS_NOV_2013.pdf, 2013.

[UNF 16] UNFPA, Earthquake in Nepal, Technical Report, United Nations Population Fund, available at: http://www.unfpa.org/emergencies/earthquake-nepal, 2016.

[UNI 15a] UNISDR, "25 April 2015 Nepal Earthquake Disaster Risk Reduction Situation Report", Technical Report, available at: https://www.unisdr.org/we/inform/publications/44170, 2015.

[UNI 15b] UNISDR, "Sendai Framework for Disaster Risk Reduction 2015–2030", available at: http://www.unisdr.org/we/coordinate/sendai-framework, 2015.

[VER 12] VERVAECK A., DANIELL J., "Japan – 366 days after the Quake... 19000 lives lost, 1.2 million buildings damaged, $574 billion", available at: http://earthquake-report.com/2012/03/10/japan-366-days-after-the-quake-19000-lives-lost-1-2-million-buildings-damaged-574-billion/, 2012.

[WIK 11] WIKIPEDIA, "Tõhoku earthquake and tsunami", available at: https://en.wikipedia.org/wiki/2011_T%C5%8Dhoku_earthquake_and_tsunami, 2011.

[WIK 16] WIKIPEDIA, "Effects of Hurricane Sandy in New York", available at: https://en.wikipedia.org/wiki/Effects_of_Hurricane_Sandy_in_New_York, 2016.

[WWJ 11] Wireless Watch, Japan, *Japan Quake* ushahidi crisis map, available at: http://wirelesswatch.jp/2011/03/13/japan-quake-ushahidi-crisis-map/, 2011.

[YAM 09] YAMINI, "Ham radio helps combat Cyclone Aila", RadioandMusic, available at: http://www.radioandmusic.com/biz/radio/ham-radio-helps-combat-cyclone-aila, 2009.

5

Context-Aware Public Safety in a Pervasive Environment

This chapter focuses on the aspects of making mobile applications and networks for public safety, especially those catering to emergency situations, context-aware. Making public safety systems context-aware would redefine how critical situations are handled today. This would enable a quick response to a situation, based on a holistic understanding of developments, as it unfolds. It is also critical that the system adapts to the limited resources and lean infrastructure available in order to function effectively.

During emergency situations, it is important to assess what information is available to us and how we deal with it. It would be of immense help if the developments in the situation were recorded and aggregated through sensors, online data sources, social networks, etc., and compiled to make meaningful use of a single

---

Chapter written by Shivsubramani KRISHNAMOORTHY, Prabaharan POORNACHANDRAN and Sujadevi VIJAYA GANGADHARAN data source. Critical decisions can be made efficiently if semantically punctuated data are at our disposal.

To make such a system feasible, it is also important that it is able to and function on an infrastructure with austere resources. During emergencies, the system's reliability would depend on how well it could overcome weak infrastructure and high network load. It would be best if the system were multimodal in nature, so that it could make use of any available mode of communication – be it visual, auditory or textual, during the emergency situation.

We support our concept with the example of an implementation – AmritaMitra (www.personalsafety.in) – an emergency response system, which links the user in distress with a public safety agency, providing them with the holistic context of the situation. This enables the emergency responder to provide efficient service based on a better understanding of the situation. The AmritaMitra system enables the user to initiate a distress call using a personal safety device through any of the multi-modal communication medium. The system revolves around our context middleware that aggregates, compiles and communicates the contextual information; empowering the emergency personnel to provide an efficient service.

5.1. Introduction

It is well recognized that context plays a significant role in all human endeavors. All decisions are based on information, which has to be interpreted in context. By making the information systems context-aware, we can develop systems that significantly improve the human capability to handle a situation's context. In this chapter, we highlight the significance of the context that wireless devices should be able to support, especially in the event of a disaster to provide much more effective support to the relief efforts. In such situations, where wireless device users need to be provided with critical information for purposes such as evacuation, relief work, etc., it is important to provide information that suits the current situation of the user while taking into account the current environment around the user.

Context plays an essential role in our ability to interpret and use information presented to us. While we, as humans, consider context all the time in all of our activities, computer systems do not usually take it into account. In fact, a computer system only takes into account the context the designer considers during the design process. As a consequence, a computer system's ability to provide relevant information to us gets restricted and we end up having to sift through high volumes of information.

It is usually a difficult task for emergency personnel to gather information about the situation when an emergency call is made. The dispatcher has to ask a lot of questions to get a precise idea about the situation, which causes delays and also may involve the exchange of inaccurate information when time is critical. AmritaMitra is a next-generation emergency response system that provides a more efficient and timely service. It is a significant advance in how emergency calls are made to systems such as 911 or its equivalent. AmritaMitra enables a person to establish an audio and video stream connection with a PSAP (Public Safety Answering Point), to give the dispatcher a precise idea about the situation. More interestingly, it also enables the multimodal communication to maximize its reach during the emergency situation by sharing the contextual information of the victim to friends and family using social media networks, such as Twitter, Facebook, etc. This enables anyone who is part of the social network and who happens to be located in the surrounding area to reach out to the victim in a timely manner. A dispatcher can also forward the stream to a responder, such as a squad car, nearest to the location of the emergency or simply the most appropriate one, to ensure a timely service. AmritaMitra is a context-aware framework, which caters to the development of mobile applications. As technology is used to augment human capabilities, it is important that the applications combine the context of the end user, the system and the environment, to provide relevant functions and the most appropriate support to users. Our framework uses a paradigm for handling context information that includes user-specific context combined with common context. The framework supports connections to mobile devices using any communication means available and connections to back-end data servers, application servers and service providers.

5.2. Context awareness

Context plays an essential role in interpreting and using information presented to us. We consider context all the time in each and every activity, but computer systems do not usually take that into account. In fact, a computer system only takes into account the context the designer considers during the design process. As a consequence, a computer system's ability to provide relevant information to us gets restricted and we end up having to sift through high volumes of information.

Consider a scenario wherein we are visiting a new city for the first time. As is the trend today, we would naturally depend on our mobile devices to make decisions on the go as we spend our time in the city. As we arrive at the airport, we may bring up Google Maps to check the distance to our hotel and the traffic situation on the way. We may use an Uber/Lyft app to book a cab. The app notifies us that the cab arrives in 14 minutes. We reach the hotel, wait for the check-in process, collect the room key and go to the room. We may pull up the WhatsApp application, check out the new messages and chat with a friend. We plan to have the friend over to go out for dinner and search for a good restaurant nearby. After having dinner, we come back to the room, set an alarm for an important meeting tomorrow and go to bed. In the middle of the night, we feel uneasy and breathless and call 911 emergency services. We explain to them our condition, medical background, etc., and wait for the service to arrive.

Let us consider the same scenario in a different way. We arrive at the airport. As we switch on the phone after landing, it automatically figures out the hotel reservation we have and brings up a map, giving us the distance, traffic information and travel time to the hotel. As we book a cab on Uber/Lyft, the phone estimates we have about 14 minutes for the cab and points out our favorite Starbucks near baggage claim carousel # 11. As we get into the cab, the phone gives us the weather information and the forecast. As we enter the hotel, we go up to a kiosk in the lobby and touch it with our phone. The kiosk gets all the required information to check us in and verifies our identity. The phone brings up a map of the hotel layout and navigates us to our room precisely through the elevator, corridor, etc. The phone detects the WiFi and automatically downloads the videos/pictures that were pending on our WhatsApp. As we chat with the friend, it understands that he is coming over and that we plan to go out for dinner. It automatically sends the friend the directions from his house all the way to the hotel and precisely up to the room. It understands that both you and the friend are vegan and lists all of the restaurants with vegan options nearby, along with customer reviews. It automatically makes a table reservation at the restaurant you select. After dinner as you are about to go to bed, the phone automatically sets an alarm considering the meeting scheduled for the next day, as well as the average travel time to the destination. In the middle of the night, as you feel uneasy and breathless, you just press a button on a special pendant that you wear and a connection with the 911 emergency service is established, your medical background is automatically provided to the personnel along with the information like what kind of food you consumed in the last few hours, etc. Moreover, the emergency service personnel does not need to ask any questions as he/she is quite aware of the situation, even before they get to the location to help.

When we compare the above two scenarios, we can see that life is much more simple in the second one. What makes it more effective is that which defines the term context awareness. In the second scenario, the mobile device seems to be quite aware of the person it is associated with and the situation that the person is in. We say that the device and the applications in the device are context-aware. From available static and live data from data sources and sensors, the system senses the situation and tries to adapt to it, mimicking a human mind.

A situation is a temporal state describing activities and relations of detected entities in the given environment [BRD 07]. A situation can be understood as a snapshot of all the possible contexts at a given point of time, with respect to an entity or application.

5.2.1. *Context*

The term context finds definitions in different domains, from the dictionary, to linguistics, to computer science. The term context has been extensively used as a concept for understanding words in linguistics and understanding the circumstances surrounding an event. Computer science has used the term context for specific circumstances, such as context switching in operating systems and context-free grammars in theories of computation, but only in the last two decades has a general definition of context attempted to be developed, specifically for context-aware computing.

In the field of context-aware computing, one the most quoted definitions of context is by [DEY 01] “Context is any information that can be used to characterize the situation of an entity. An entity is a person, place, or object that is considered relevant to the interaction between a user and an application, including the user and applications themselves”. The focus here is on adaptiveness of the application rather than change in its behavior.

Context is essentially a property or an attribute of any of the elements of the pervasive system. It can also be the ability of an entity in the system to carry out an activity or the role that the entity plays in an activity. It can have a type and description, and a value that can be discrete/nominal. It can also have a hierarchical structure. With this perspective, location and identity can be considered as properties of an entity and hence part of its context. Consider a person with motor disability, this piece of information is part of his/her context and can play an important role in situations that require him/her to perform certain tasks which require movement.

We do not attempt to refine the term context, but try to extend the definition taking into account the practical implementation of applications that are context-aware. Our definition, as found in [CHR 10], is:

Context consists of one or more relationships an information item has to another information item. An information item can be any entity, either physical (such as a person, a computer, an object), virtual (such as a computer service, a group of people, a message), or a concept (location, time, and so on). A relationship describes a predicate connecting two or more information items, which may change at any time for any reason.

5.3. Context-aware middleware

Dey *et al.* [DEY 01] state – "A system is context-aware if it uses context to provide relevant information and/or services to the user, where relevancy depends on the user's task". They focus on adaptiveness of the application rather than change in its behavior. They also specify certain features that a context-aware application should support: presentation of information and services to a user, automatic execution of a service for a user and tagging of context information to support later retrieval. A context-aware middleware acts as an integration point that compiles the context for the given situation, enabling the applications, subscribed to it, to be context-aware.

Middleware is considered an indispensable component of context-aware environments. Ranganathan *et al.* [RAN 03] argue that ubiquitous computing environments must provide support for middleware. This is because the middleware would provide uniform abstractions and reliable services for common operations and would simplify the development of context-aware applications. It would be agnostic to hardware, operating systems and programming language. It would also allow us to compose complex systems based on the interactions between a number of heterogeneous and distributed context-aware applications. More importantly, it would provide support for complex tasks, such as the acquisition of contextual information, reasoning about context using mechanisms (e.g. rule based) or temporal or spatial reasoning as well as learning from context using mechanisms like Bayesian networks, neural networks, reinforcement, supervised and unsupervised learning and modifying its behavior based on the current context. It would also define a common model of context, which ensures that different applications in the ubiquitous environment have a common semantic understanding of contextual information. They also specify certain requirements for middleware for context-aware systems in ubiquitous environments, which in today's terms mean:

1) It should support collection of context information from heterogeneous sensors and services as well as the delivery of appropriate context information to different applications;

2) It should support inference of higher level contexts from low-level sensed contexts;

3) It should provide tools for different kinds of reasoning and learning mechanisms;

4) It should allow applications to easily behave differently in different contexts;

5) It should enable syntactic and semantic interoperability between different applications and services.

5.4. Practical experience – implementation of AmritaMitra personal safety framework

The system can be accessed by web-based portals from the Internet at www.personalsafety.in. The personal safety framework consists of two major subsystems which are as follows:

a) sensing and alerting;

b) cloud services and contextual analysis.

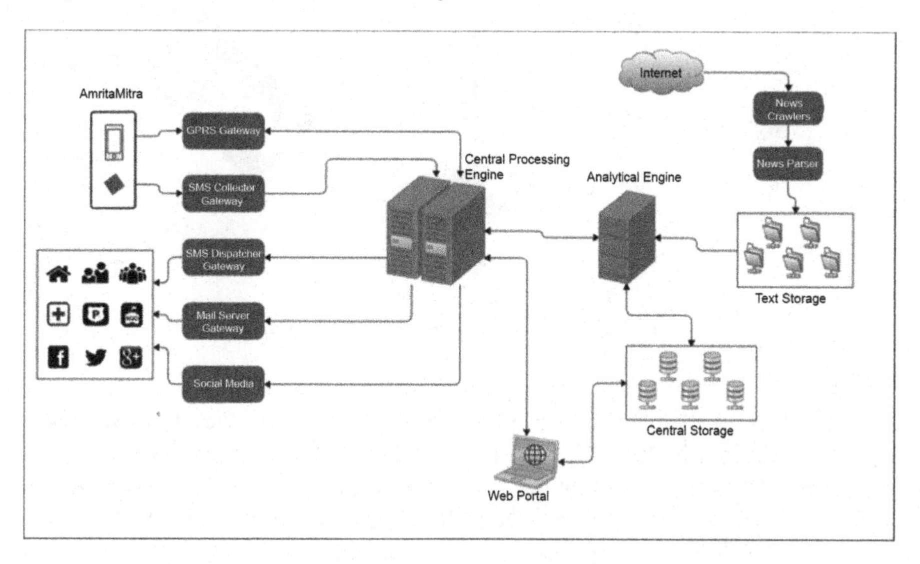

Figure 5.1. *AmritaMitra context-aware framework architecture*

5.4.1. *Sensing and alerting*

Sensing is performed by a smart phone application (available in Android and IOS platforms) or by a dedicated wearable device. Both the devices use microphone, accelerometer, GPS and camera sensors. The sensing devices (smartphones or wearable personal safety device) gather information about the current user location and their surrounding information and feed into a decision-tree-based system that computes the degree of safety of the user.

The request for emergency help can be done by automated triggering mechanisms or by the user pressing the SOS button manually to summon for help. The section below describes the various modes available.

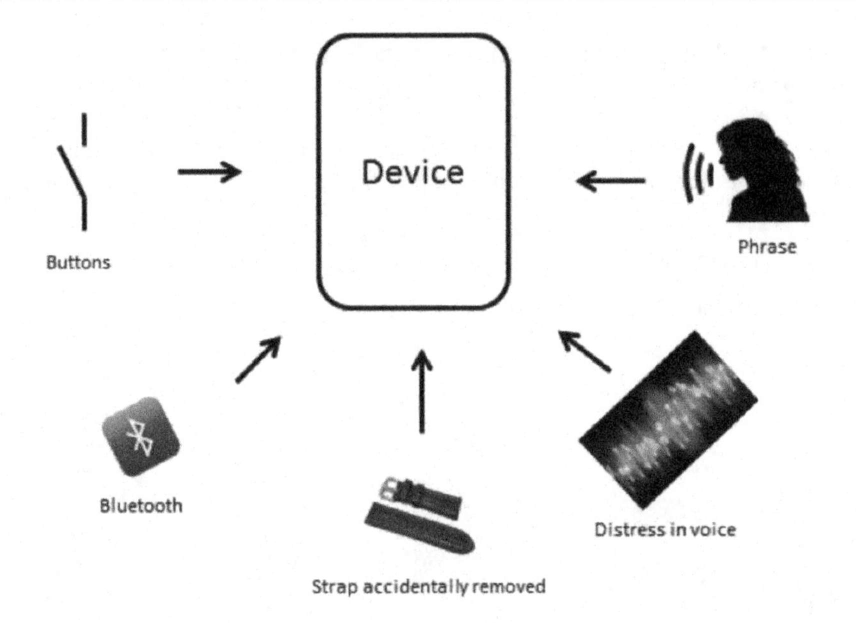

Figure 5.2. *Modes for triggering an alert*

Voice-based activation: as shown in the diagram above, the emergency help request is triggered by uttering some phrases; for example, sentences like "I need help" can be used as a triggering mechanism. The system can also be configured to detect highly distressed voices/loud shouts to trigger the emergency mode. In certain emergency situations, the adversary could remove the wearable device from the victim forcefully, and this can be configured for use as one of the methods for triggering the emergency mode.

With the advent of Bluetooth Low Energy (LE), which consumes much less energy for always-on applications, Bluetooth is also harnessed as one of the methods for initiating the alert. A Bluetooth LE-based device that is worn as an accessory achieves this. This is primarily useful in a situation where the victim cannot reach out to the smartphone in the emergency situation, or when the smart phone is snatched from the victim.

The triggering of the alert creates a GPRS message with the user location and the status of the information and sends it to the cloud. When the GPRS and other data channels are not available, SMS is used as a fallback mechanism to send the user state information to the centralized Cloud-based Contextual Analysis.

5.4.2. *Cloud services and contextual analysis*

The cloud services engine performs two main functions. First, it aggregates the contextual data about a location that are fed by data crawlers. The data crawlers are programs that traverse several media web portals and public databases to collate information on all the crime incidents, including robbery, theft, and violence against women, children and other vulnerable citizens. Second, it also receives the SOS information and routes emergency information to the responders and social circle according to the rule engine. The management of the user information and the emergency contact management is performed using a Web-based interface and API.

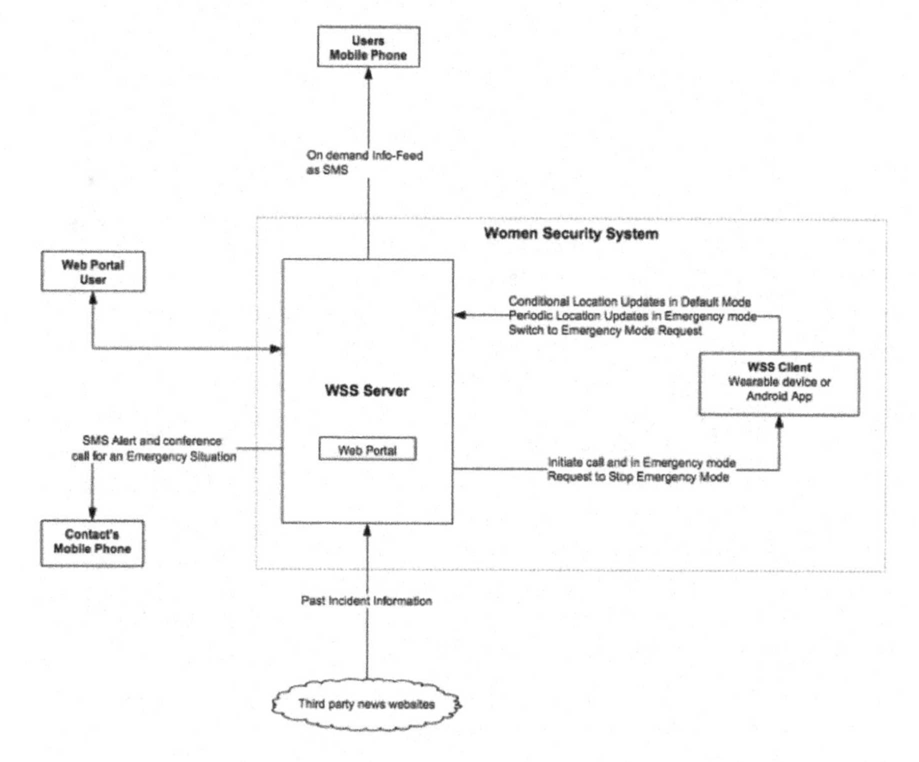

Figure 5.3. *Workflow of the emergency communication system*

A dedicated crisis management web portal acts as a primary interface to manage user information, contact information, contextual logic rules, etc. When the AmritaMitra smartphone application is installed on the user's mobile for the first time, the application provides a facility to register user information. Once the account is created, the user is provided with the credentials to log in and use the web portal, which is created for managing the personal safety of the individual. In addition to the standard emergency contact lists, the user can also add or edit the contact information of their friends and relatives who wish to be contacted in case of an emergency. The web portal can be used for monitoring and tracking the victim in real time. The web portal contains the real-time location of the user marked on a map. The map also shows the availability of nearby friends, police stations, NGOs and hospitals and their respective distance from the user. This web portal can also be used to monitor the places where any alert request is made in real time. The map also shows the number of attacks and instances of physical violence, which occurred in the past.

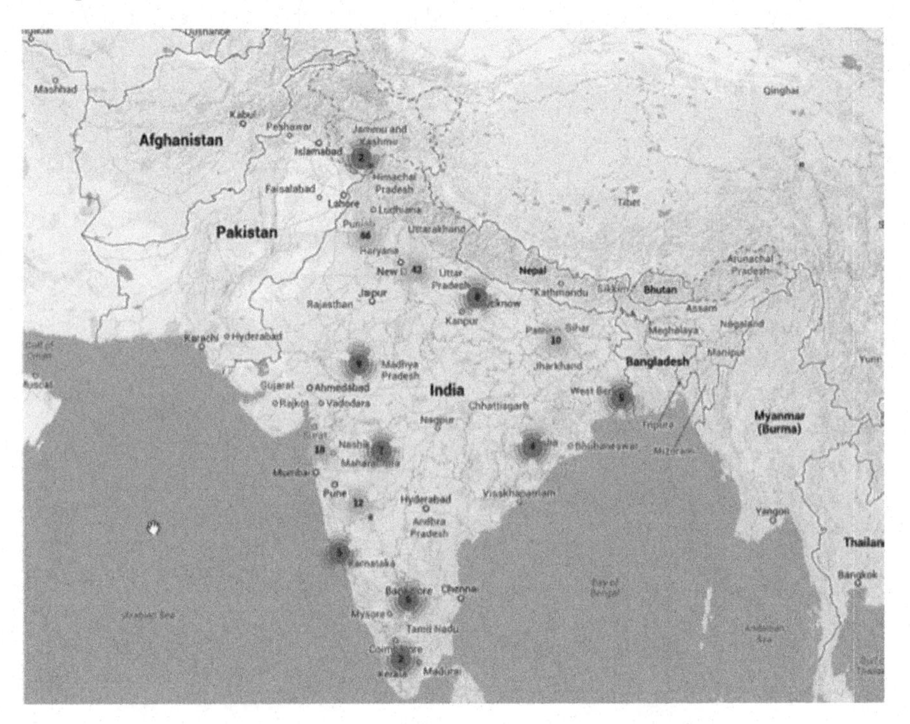

Figure 5.4. *Map showing the locations of reported violence against women in the month of December 2015*

5.4.3. *Social media as an emergency communication system*

The ubiquitous nature of social media has led to it becoming an important communication medium. There is a large amount of contextual information that can be obtained by social media, including user location, temporal information, event occurrence, presence of friends in the nearby locations, etc. AmritaMitra framework uses social media networks as an important communication media for summoning help as it leverages all the above-mentioned contextual information. In the event of any violence or physical altercations, the framework reports the location and the nature of the incident to all the friends in nearby locations, thus enabling them to help the victim much faster than the law enforcement. The social media networks that are currently integrated with the framework are Twitter, Facebook and Google+.

5.4.4. *Indoor and outdoor location identification*

The AmritaMitra safety device has a GPS module through which location details are obtained in latitude and longitude format from satellites. Usually, the GPS fix takes 40–60 seconds but it can take longer when cloudy or with other bad weather conditions. Hence, we have implemented two additional methods to speed up the location identification when the GPS fix takes time. These are the mapping of the CellID and the GSM cell triangulation method. The GSM communication component provides CellID information.

The CellID is the unique ID allotted to a hexagonal area used by mobile towers for location identification purposes. These data, which are acquired by the smartphone or the wearable device, are parsed and used to identify the corresponding coordinates of the location. A second method known as cell triangulation is implemented based on the location coordinate data of the mobile towers available for access.

5.4.5. *Real-time user and context tracking*

The framework tracks the victims in real time until the situation is under control and the first responder and the victim or authorized relative agrees to stop the tracking. The real-time user tracking is very useful in cases of abduction or any other situation where the victim is on the move. Location tracking is achieved by using GPS sensors and cellular networks identification described below.

The portal uses Google maps for displaying the real-time location of the user. By default, the map is zoomed in to a high level, so that it is very convenient for people to track and monitor a user. The map is refreshed and updated every 5 seconds for proper monitoring. The map also shows the nearest police stations, hospitals and a list of friends whose geo-location is available along with the distance and direction from them. Figure 5.5 gives an example in which a friend of Bob is tracking him. The tracking person gets to see the exact location of Bob, his current activity and the speed with which he is moving. On the map, we can also see nearby police stations, hospitals, and online friends and the distance from them.

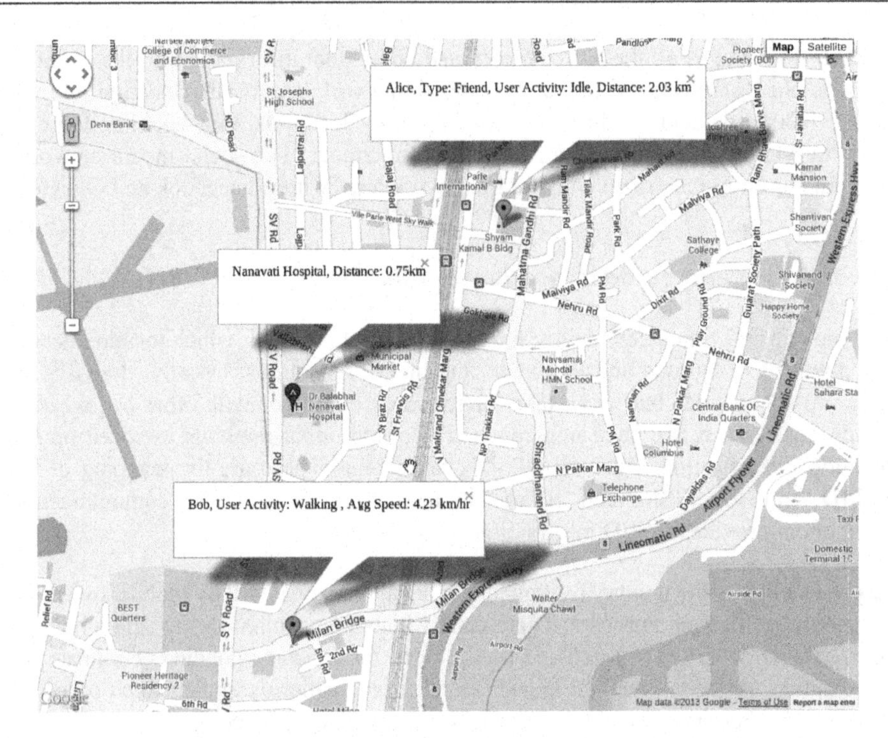

Figure 5.5. *Tracking Bob in real time using the AmritaMitra framework*

5.5. Conclusion and future directions

In this chapter, we discussed the importance of context awareness when dealing with emergency situations. In a world where we are almost always dependent on mobile devices, it would be best that the device not just assists us but does it with a good understanding about the user and the situation the user is in. It would be particularly useful in a critical situation.

We discussed in this chapter the importance of context awareness and context-aware applications and systems. We highlighted how a context-aware pervasive environment could be built around a middleware, where contextual information is sensed, aggregated, compiled and utilized according to the situation. A suite of applications, subscribed to the middleware, can create a context-aware ecosystem based on sharing of context between applications. We exemplified the significance of a context-aware ecosystem, especially in an emergency situation, through AmritaMitra – a context-aware emergency response system that helps the user reach out to a 911-like system just by a click of a button on a small physical device. The mere press of the button communicates to the emergency personnel contextual information like the user's background information and current situational information along with an audio and video communication stream.

5.6. Bibliography

[BRD 07] BRDICZKA O., CROWLEY J.L., REIGNIER P., "Learning situation models for providing context-aware services", *Proceedings of the 4th International Conference on Universal Access in Human-Computer Interaction: Ambient Interaction,* UAHCI'07, Springer-Verlag, Berlin, Heidelberg, pp. 23–32, 2007.

[CHR 10] ALMAZAN C.B., ROVER, "Architectural support for exposing and using context", PhD Thesis, University of Maryland, College Park, 2010.

[DEY 01] DEY A.K., "Understanding and using context", *Personal and Ubiquitous Computing*, vol. 5, no. 1, pp. 4–7, 2001.

[RAN 03] RANGANATHAN A., CAMPBELL R.H., "A middleware for context-aware agents in ubiquitous computing environments", *Proceedings of the ACM/IFIP/USENIX 2003 International Conference on Middleware*, Springer-Verlag, New York, pp. 143–161, 2003.

6

Supporting New Application and Services over LTE Public Safety Networks

This chapter provides an overview of different scenarios that can be identified in emergency situations, such as natural disasters or accidents, thus providing an overview of new services that might be beneficial for these scenarios as well as the major challenges that have to be addressed by public safety LTE networks to maintain communications in these situations. Additionally, a case study about a new video hardware platform to be used by first responders is introduced alongside new components to be included in the operators' networks to provide these new services in the area of public safety.

Chapter written by César Agusto GARCÍA-PÉREZ, Almudena DÍAZ-ZAYAS, Álvaro RÍOS, Pedro MERINO, Kostas KATSALIS and Donal MORRIS.

6.1. Introduction

The use of mobile broadband access has been researched intensively in recent years to meet public safety requirements and mission critical communications. This trend has now been accentuated by the appearance of new 5G technologies as they promise a major increase in reliability and performance. Technologies that are already available on the market like WiFi and LTE are being evolved to match strict requirements and fit into the 5G landscape together with new radio access technologies. LTE is the most widespread; it has now been rolled out in many countries and its standards are constantly evolving to add new features that could be used to support public safety services.

Public safety use cases span many different vertical industries and scenarios, and can include public protection services or communications in emergency situations such as accidents or disasters. Different and divergent requirements exist for supporting communication systems, the network, applications and end users, while rich features at the application level are related to key performance indicators and possibly require new functionalities in the network. Some of the communication requirements that must be fulfilled by public safety networks are:

– ability to provide broadcast/multicast communications, which can be used not only to deliver alerts but also for communication between groups of users (e.g. blue light services);

– ability to provide emergency communications; the network should be able to allow access to emergency services independent of the operator (e.g. access to emergency numbers such as 112 and 911);

– ability to prioritize traffic and provide service differentiation, especially in emergency situations in which the network tends to be congested;

– ability to maintain service levels to support voice or video communications with efficient resource management and guarantee a certain level of quality of service;

– ability to position and sometimes to address based on the position. This functionality can be used to provide communication between first responders and to locate people;

– ability to provide high reliability and availability. A public safety system should be available for everybody, independent of the carrier, and should be reliable even when using unreliable communication channels;

– ability to support new application requirements, such as moving towards 5G, to provide more services such as video (video surveillance, support for field personnel in accidents, robotic access to dangerous for life spaces, etc.) and data management (to support augmented reality, transfer of maps/plans, positioning updates, robotically controlled plane, etc.).

This chapter describes some of the background motivation and challenges that modern mobile networks have to address to provide these kinds of services, and a possible approach for some of the emergency scenarios is studied. A new system called BlueEye [GAR 16a] is presented as a commercial solution for first responders in an emergency, as a use case, and a demonstration of the feasibility of the solution is also provided. The technical approach is verified based on real deployments and experimentation platforms with a focus on providing realistic results close to those that can be obtained in real networks, which is the main distinguishing feature of this chapter, relative to numerous other research initiatives in the same area which tend to be based on simulations instead of using real world equipment. The platform, which is described in the following sections, combines the programmability that is normally found in a laboratory environment with realistic results obtained by using a commercial, albeit controllable, network.

The rest of the chapter is organized as follows. Section 1.2 provides an overview of the background, including the related work and the motivation for public safety. Section 1.3 describes the challenges to overcome in order to provide services as well as some typical scenarios for public safety. Section 1.4 presents a technical approach to overcome some of the challenges of public safety services, and finally section 1.5 presents conclusions and describes future work.

6.2. Motivation and background information

This section provides some details on the motivation behind supporting public safety services with mainstream mobile networks, the related work on public safety communications and standardization efforts with respect to the features required by such networks.

6.2.1. *Motivation*

Currently, public safety communications mainly rely on the use of Private Mobile Radio (PMR) solutions such as European Trunked Radio (TETRA), which were designed and oriented with the public safety market in mind in the 1990s. However, such networks lack the necessary capabilities to support new high bandwidth services that are required for 5G. These technologies have been designed from a voice-centric perspective while there is a clear trend towards the provision of data-centric services. Furthermore, these solutions are based on proprietary dedicated networks, normally deployed by governments and operated by contractors, and require large investments in terms of both equipment and maintenance. Mobile broadband public safety can help reduce these costs by introducing economies of scale [FER 13] and can accelerate the introduction of new functionality in the network as the mainstream systems are constantly evolving. An analysis of the economic feasibility of providing mobile broadband public safety has been provided in [PEL 15] in terms of Capital Expenditures (CAPEX), Operating Expenditures (OPEX) and the type of deployment. On the technical side, there are some characteristics in current mobile networks that can be improved based on design paradigms introduced by NGMN for 5G. The main challenges that are common for many public safety scenarios are:

– need to provide mechanisms so that public safety applications can dynamically demand resource reservations from the network;

– reduction of network latency, especially when the communication is established with peers geographically close to each other;

– improvement in the behavior of applications when used in high mobility scenarios that constantly trigger handovers;

– improvement in the behavior of base stations' schedulers by providing information on the type and importance of information transported;

– ability to provide traffic differentiation in the network, guaranteeing the delivery of the most critical data;

– new methodologies are required for selecting the most appropriate equipment for a given public safety application.

6.2.2. *Related work*

Efforts and investment focusing on investigating, evaluating and standardizing the use of LTE for Public Protection and Disaster Relief (PPDR) communications and other mission critical systems are ongoing in many countries [GER 13, AUS 15]. The principal investor has been FirstNet in the USA, backed by $7 billion dollars, to deploy an LTE nationwide public safety network [MOO 16].

An economic perspective on the provision of public safety services has been provided in [GAR 16a, FER 13, FAN 16]. The use of commercial mainstream technologies can introduce many benefits to public safety networks, although the type of deployment is still an open discussion. For instance, according to a study by the European Commission [FOR 14], shared infrastructures are only considered in the medium and long term, as the mobile network business model needs to be changed, so the requirements of public safety communication could be fulfilled on a shared infrastructure.

Furthermore, there are many approaches that cover the provision of mission critical/public systems, most of them covering specific application domains, such as [DIA 14], which explores its use for railway control communications, or [CIC 16], which explores the provision of mHealth emergency services over LTE. Some cover the provision of group communications in these types of scenarios, for instance [PRA 16], which explores several deployment alternatives for LTE-A aiming to optimize parameters such as power consumption and resource allocation.

Most of the work on SDN concerns the provision of general services. There are a few mission critical services worth mentioning, but most of the time, the approach is more general. For instance, [BOU 14] explores the use of a distributed SDN control plane able to provide communication between controllers of different domains in order to provide end-to-end QoS to services. An architecture for 5G SDN networks has been proposed in [GUE 14]; the functionality currently present in the core network is divided into different components that are managed by an orchestration controller. The authors provide some estimates on the benefits introduced by the architecture. The use of WiFi is not normally considered in mission critical communications, as the QoS cannot be guaranteed, but it might be a solution to provide coverage in indoor settings with poor mobile scenarios.

As we will argue in the following sections, Q4Health proposes a unified SDN-based solution for a public safety application where paramedics communicate with command centers/hospitals by exploiting LTE networks together with untrusted WiFi access points. These are integrated using the non-3GPP access functionality to support a seamless handover from or to LTE networks.

6.3. Services for public safety networks

The study of real-time video transmission in an emergency situation makes sense from the viewpoint of network optimization, as it is one of the most demanding services that can be used. The high bandwidth needed to transmit and receive the video without stuttering and the low latency requirements in a live transmission stress network capabilities differently to other services, and any development towards ensuring this type of communication can therefore easily be translated to other, less demanding types of services.

Different events or emergencies require different approaches and resources available in the network. This section tries to enumerate the common challenges for network operators in different emergencies and then the particular characteristics of some types of public safety scenario, discussing the areas that should be improved in the network infrastructure. Then, the main use case of this study and the BlueEye platform are presented.

6.3.1. *Common challenges for current technologies in emergencies*

Although, at a first glance, existing Radio Access Technologies (RAT) like LTE and WiFi, or a combination of both, can be used to cover the communication needs of the emergency personnel in such events, a number of optimizations need to be applied to ensure the requirements of public safety systems, in particular traffic prioritization and QoS enforcement in the network, voice and video transport, multicast communications, and so forth. In order to support uninterrupted communication with specific QoS requirements, the main shortcomings identified in existing RAT that are commercially deployed are as follows:

– an inability of the application to request from the mobile network a guaranteed bandwidth or a quality of service profile. Even though standard interfaces for requesting such types of services exist in the international specification of radio systems, in practice, they are not available today for a number of reasons, including technical, operational, governance and commercial ones;

– in our demanding case study, a staggering 5 s end-to-end delay was measured for the interactive simplex video and duplex audio, compromising the ability to have a fluid and streamlined conversation between two parties, for example a paramedic in the ambulance and the medical team at the hospital.

Any solution for these challenges should also include mechanisms to produce Charging Data Records (CDRs) for commercial networks or risk facing resistance to the introduction of such services from the operators. Furthermore, one of the most important challenges to consider in the provision of public safety services is the robustness of the network. There have been several studies that analyze the possible attacks on LTE networks. For instance, [LAB 16] describes techniques to reduce LTE spoofing attacks (introduction of an intermediate cell broadcasting in the control channels to produce denial of service). In [LIC 16] and [AZI 15], jamming attacks that consist of generating a high power signal to interfere with control channels have also been described. Development to provide new services in the network should also take into account these shortcomings, especially given the sensitive nature of the public safety agency traffic.

It also seems that one appealing scenario for governments is the combination of dedicated public safety network deployments with the sharing of infrastructure with commercial mobile networks. For instance, dense urban areas can be supported with the former while rural areas with the latter. To support infrastructure sharing, more work on service isolation has to be done so as to ensure no between-network effects. The random access procedures also have to be improved as preamble collision will tend to become more frequent not only due to machine-type traffic but also because of the increase in traffic due to emergency situations.

6.3.2. *Specialized scenarios in public emergencies*

This section describes some additional caveats that public safety communications could face in particular scenarios.

6.3.2.1. *Events with big public responder's deployment*

In these scenarios, we include anything ranging from industrial accidents to terrorist attacks or riots, situations where a fast deployment of emergency services is expected just after the event. In this case, video communications, which are the model for our study, are predominantly upward, i.e. from the end user, be it a police officer or firefighter, to a central location or headquarter where these videos are aggregated and processed. From the viewpoint of the network, these scenarios are defined by a sudden increase in the utilization of the network, usually due to bystanders near the incident, and then by the addition of the emergency services traffic on top of an already overloaded network.

This type of scenario imposes two burdens on the network. On the one hand, the resources available for the uplink channel are limited by the real capacity deployed in the area, because usually it is the downlink that is optimized as it is the usual direction of the network traffic. On the other hand, the network should be able to identify and process the traffic coming from public safety personnel with a higher priority than the rest of the users, maintaining a best-effort policy for these or even reducing or disabling the non-priority traffic if the priority one needs more bandwidth.

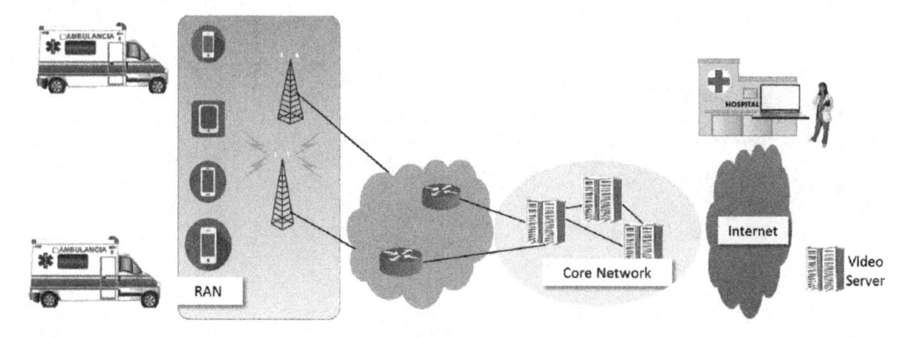

Figure 6.1. *Scenario with public responders*

Figure 6.1 shows a common first responder with all the elements of the LTE network. Typically, the communications required will be multicast, and the peers of the communication might be in high mobility (vehicles), low mobility (blue light personnel on site) and no mobility (hospitals, command centers, etc.). As mentioned previously, the first responder communications will normally have to compete with the rest of the network traffic that has to fulfill the quality of service and prioritization.

In addition to the technological challenges associated with identifying and interconnecting the networks of the different operators available in the emergency area, cooperation between political and industrial partners is needed to define the events in which those services are activated and to decide who is the final party responsible for coordinating safety assets.

6.3.2.2. *Natural disasters*

Earthquakes, flooding or large hailstorms have the same traffic pattern as the previous scenario, with an explosion of civil traffic moments after the event, and the subsequent addition of emergency services data that should be prioritized over any other type of communication. In addition to these problems, these types of events add a new challenge to the network as a natural disaster will usually reduce the physical resources available to the network. Cell towers are no longer aligned with their corresponding radio link, or even fall from the top of buildings, and blackouts are normal in these situations; moreover, it is not unusual for public services to ask citizens to turn off their devices to make room for the emergency traffic.

Figure 6.2. *Natural disaster scenario*

Figure 6.2 shows a natural disaster scenario; in these types of scenarios communications might be physically damaged so it might be necessary to have a fast deployment of the infrastructure used to support communications or use satellite communications if the applications are susceptible to delays. For instance, the use of unmanned vehicles can be useful to get information from zones that are not accessible (e.g. due to radioactivity, fire, etc.), but the communications will have to provide it with a dedicated deployment.

Nevertheless, the kind of deployment that a network operator needs to carry out to guarantee the service in this type of scenario is the same as any other high-resilience network, which has been studied in depth previously [EUR]. This means that a real-time high priority service for emergencies, the focus of this chapter, is built on top of an already functioning network so the kind of deployment required is ensured to be the same as for other events.

6.3.2.3. *Emergency broadcast*

Video broadcasting with safety information to users is expected to be enabled by the increase in bandwidth, which new technologies like 5G will provide. There are already systems, such as Wireless Emergency Alerts in the United States or the European Public Warning System in Europe [WIR, ETS], that use the network's broadcast channel to provide highly localized security information to end users, but they only use low bandwidth SMS-like messages in the communication.

The appearance of video messages for this type of communication is not expected to cause additional challenges to the network aside from the normal evolution of the technology as, in this case, the operator controls all elements in the stack and, if needed, can increase or unilaterally reduce the priority of all other traffic to ensure these messages are received. Also, broadcasting of this kind is likely to be unidirectional from the network to the user and, without the possibility of a response, latency requirements for providing real-time communication are greatly reduced.

6.3.2.4. *Localized emergency services*

Although the scenarios described in the previous subsections are serious and challenging to the capabilities of the network, they are also very rare. The focus of the Q4Health project is to enable small groups of users to use network resources with a guarantee of service, or at least with a higher priority than commercial services, in situations where the rest of the network works as expected. The use case that the project is built around is one where an ambulance responds to a medical emergency while sending audio and video to the hospital where a medical team can recommend different treatments, which is explained in detail in the following section. It is, however, just one of the many scenarios where first responders need to provide a visual characterization of an emergency to a remote location.

In contrast to the previous scenarios in these events, there is no network shortage and the traffic of first responders in an emergency has to be processed alongside the traffic of normal users, albeit with higher priorities. The overview of the BlueEye hardware platform detailed in the next section is accompanied by the introduction of SDN components in the core network to enable the required prioritization while causing minimal disruption to the network operation. It also includes novel communication services, like group communication, without the requirement of specialized equipment from neither the end users nor the operator's network.

6.4. Wearable devices in public safety

Up until recently, sending video back to a remote location in an emergency situation required an additional first responder tasked exclusively with the operation of the camera and associated transmission devices. The evolution of mobile technology has allowed the miniaturization and reduced consumption requirements of such devices to the point of making high-definition cameras available in a low-cost mobile phone. However, this capability is still insufficient as it would be cumbersome for a first responder to operate a mobile phone while helping the victims of an accident, for example. The apparition of wearable computing devices like smartwatches has opened the door to new forms of real-time communication for such professionals, allowing them to receive texts or maintain a bidirectional conversation without limiting their mobility or ability to use the hands. Some countries have begun the deployment of wearable cameras to police officers and firefighters, but there is still room for improvement as these devices or their batteries are still too heavy and tend to be chest-mounted instead of head-mounted, which would be preferable so that the headquarter or hospital receiving the video would know exactly what is the field of vision of the first responder. One of such devices is presented in the next section.

6.4.1. *The BlueEye use case*

The use case that we consider in this section shows the way in which immersive wearable live video can assist first responders in public safety applications where an ambulance paramedic is attending to a patient. Generally, the paramedic's role includes autonomous decision-making on the emergency care of patients. It greatly depends on the paramedic's ability and responsibility to successfully handle the incident and provide advanced treatments in the field, rather than simply transport patients to the hospital. In the "golden hour" after the incident, pre-hospital actions taken are crucial and potentially lifesaving. "Eyes on" live video from the first responder to the hospital or clinical oversight center can help make more rapid decisions.

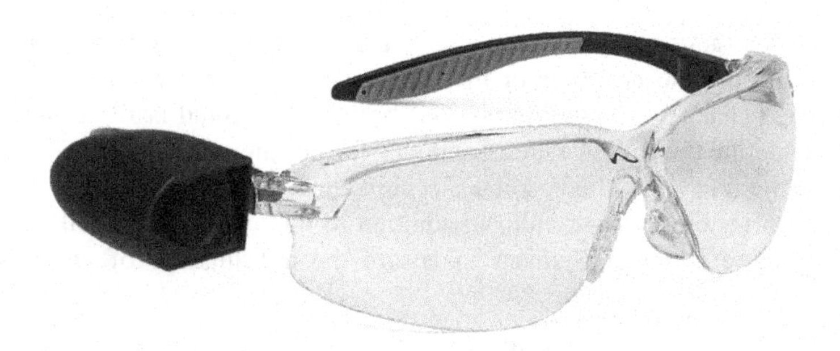

Figure 6.3. *BlueEye wearable device*

The BlueEye platform [WEE 12] consists of head-mounted camera glasses that can be used by first responders to send video in real time to hospitals and command centers coupled with an embedded computer for processing and different radio technologies like LTE or Wi-Fi to connect to the network. The BlueEye system was developed by RedZinc, Ireland, as an interactive wearable video platform for ambulances and paramedics.

6.4.2. *BlueEye for first responders*

As mentioned above, the use case studied is the support of a paramedic provided with COTS equipment to provide a simplex video and duplex audio transmission to enable emergency doctors at the hospital to remotely see a pre-hospital patient experiencing an acute medical emergency and provide additional treatment possibilities for that patient. During the "golden hour", rapid diagnosis and early treatment have the potential to improve a patient's outcome (e.g. more rapid delivery of clot-busting drugs).

The utilization of COTS equipment is preferred to counter the use of specialized devices such as those currently required if TETRA or any other proprietary technology is used. The equipment used is described in more detail in the next section. Current technology deployed nowadays also has a series of identified shortcomings that need to be addressed:

– There is no mechanism for the UE to select priority scheduling at the critical moment when the paramedic arrives on the scene and so an emergency video connection of typically 10 min duration shares resource blocks with "run of the mill" traffic including Facebook, YouTube and other smartphone traffic. As the use of COTS devices is envisioned, there is a need to find a method to signal the network so that all or part of the traffic coming from the equipment of public responders should be provided with extra guarantees and safeguards.

– When the paramedic enters a house or a building, the video signal can be lost due to a drop in radio coverage or the appearance of a more powerful signal that triggers a cell change. Despite the obvious benefits this could bring to the society, there is no defined seamless video handover for any new cell to enable backhaul over the public Internet to the hospital. Technologies like WiFi offloading have been designed to offer higher bandwidth in addition to the usual LTE mobile data rate, but, again, there is no mechanism to ensure the transmission of video without breaks.

– In certain situations, it would be helpful to patch the video using nearby supervising vehicles to link the paramedic and the hospital for additional support in an emergency event. However, with the current implementation of mobile technology, this implies a trombone from the source eNodeB to a centralized EPC and back to the target eNodeB, from where the signal has to be broadcast. One way to add this capability to the network, using an SDN approach to route traffic without the cooperation of the EPC, is being studied and will be tested with the main use case.

– A significant growth in IoT eHealth monitoring sensors (e.g. ECG, pulse oximeter) is anticipated with various demand profiles on the network, and it is prudent to fine tune access capability to accommodate these eHealth sensors, rather than blending them with every traffic source in a best effort.

6.4.3. *VELOX API for public safety applications*

In this section, we describe the proposed solutions to the shortcomings described above and the technical approach used to validate the results, providing an overview of the architecture currently being developed as well as details on the most relevant aspects to optimize the network.

6.4.3.1. *Proposed architecture*

Figure 6.4 shows the proposed architecture integrated within the LTE stack. In the radio access, the use of SDR solutions is foreseen so as to validate the application-aware MAC schedulers and also to provide more information regarding the performance of antennas. Commercial base stations will also be used in the scenarios to validate core network enhancements and will be combined with WiFi to validate the seamless offloading of traffic. New components to support latency reduction are foreseen such as the *Fog Gateway* (F-GW), a computing component based on the analysis of the base station data plane, or the *Middlebox* which provides a similar functionality but analyzes the control plane. The rest of the components are the standard ones provided by a commercial deployment.

Additionally, an evolved Packet Data Gateway (ePDG) in charge of supporting the WiFi offloading is also introduced. All these components (base stations, latency reduction elements and core network) are connected by one or more instances of Open vSwitch (OVS) emulating an SDN backhaul. To add more control and measurement points to the core components, the radio equipment and the user subnetwork are connected with a multi-domain distribution network, emulating the transport among different domains of traffic to the servers.

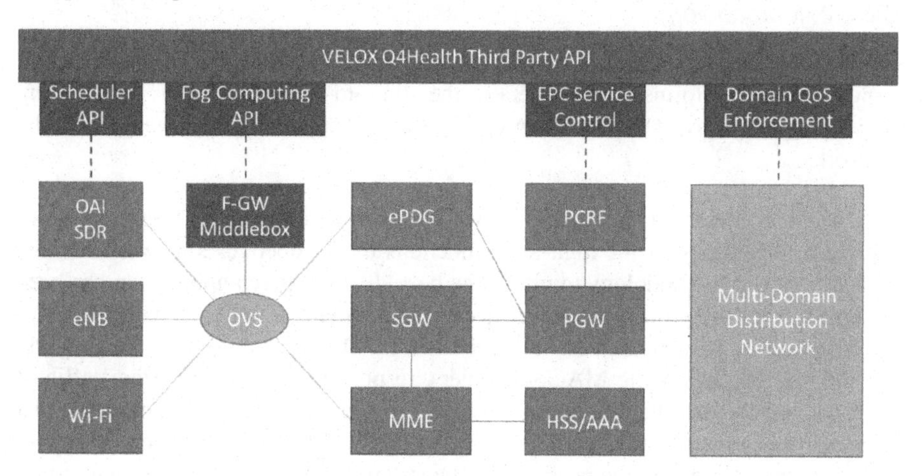

Figure 6.4. *VELOX network architecture*

The central role in the architecture is the software-defined network platform VELOX. VELOX is a multi-domain orchestrator that offers an API for third party applications to invoke priority bandwidth on the network. It has a modular design with drivers to interact with different components, and has already been extended to provide the following functionality for our case study:

– provision of the type of application that a user is using in the schedulers of the base station in order to optimize resource allocation decisions;

– interfacing with latency reduction elements, which involves setting up the SDN backhaul to forward the packets and enabling the gateways functionality;

– support of QoS demands to the core network. As operators are normally reluctant to offer access to the standard QoS interface of the EPC, the VELOX driver hides and wraps the functionality, avoiding the exposure of the API and adding an additional security layer;

– QoS enforcement across multiple domains. In the future, transport networks may offer access to functionality to support QoS enforcement across different paths.

This component negotiates with SDN switches of each domain in order to guarantee a certain level of quality.

This third party API provided by the VELOX engine can be used to integrate applications more easily, and offers a single configuration point that can be used to improve the behavior in all the components of the network.

6.4.3.2. *Methodology*

The methodology consists of executing a set of experiments on two different experimental platforms to accelerate the time-to-market of the system (by optimizing its behavior) as well as preparing for the upcoming improvements in the networks.

The experiments are designed to cover all the components of the network, from the radio technology to the high-level functions of the operator's network. On the end-user side, a methodology to match the best UE for a given application has been designed, from the processing power needed to maintain a high bandwidth video to the selection of the antenna best suited to the task. On the radio side of the network, the focus is on optimizing MAC schedulers, responsible for planning the traffic of all users, which can be improved by providing information on the type of traffic that is being transported. The experiments on the EPC explore the reduction of system latency, especially when the peers of the communication are geographically close to each other, and the provision of efficient mechanisms to support multicast communications.

The experimental platforms used to validate the proposed architecture are PerformNetworks [DIA 15] and OpenAirInterface [KAL 15]. PerformNetworks is a dedicated testbed for mobile communications, which provides multiple components including commercial base stations, core network emulators or conformance testing equipment. The components can be interconnected and mixed in order to test hybrid scenarios with real equipment, which is useful not only for obtaining realistic results but also for controlling the experiments and providing highly detailed measurements while maintaining the reproducibility. OpenAirInterface (OAI) is a Software Defined Radio (SDR) of LTE and 5G network stacks. The platform provides support for the UE, the base station and the core network, and it is compatible with commercial equipment, which enables the evaluation of new features on the network with standard components. The platform has been used to support many features such as broadcast communications and cloud Radio Access Network (RAN).

6.4.3.3. *QoS enforcement*

To satisfy the QoS requirements of public safety applications across all the components of the network, several strategies are followed.

On the one hand, the standard QoS interface available only to the operators of the core network, the Rx interface, is wrapped and exposed to third parties so they can ask for a determined bandwidth as well as identify the type of traffic (real time or streamed). With this information, the EPC Service Control driver sends a demand to the Rx interface asking for the provision of the desired bandwidth or priority. If there are sufficient resources in the network to process the request, one or more dedicated bearers will be established as a result of this request. The dedicated bearer provides QoS not only in the components of the core network, but also in the radio access; thus, it will prioritize the packets in the MAC layer, also considering the type of bearer in which they are to be transported.

The API should provide support to enforce QoS across multiple domains, setting up the SDN switches in the distribution network to change the traffic from best effort to expedited forwarding. Although nowadays it is not very common to have these types of access for switching the distribution networks, which are frequently owned and operated by contractors, the authors expect that this situation will change in future 5G networks as it will be necessary to fulfill advanced application requirements.

Finally, a version of OAI eNodeB has been modified to offer a programmatic API for the MAC scheduler, the objective of which is to apply SDN principles to radio access, providing a view of lower layer resources to a centralized element and also generating schedulers that are aware of the type of traffic that they are transporting.

With these three features, the QoS is enforced end to end in each of the components of the network and is exposed to third party applications that can dynamically trigger Service Level Agreements on the network.

6.4.3.4. *Latency reduction techniques*

Network latency can be a very important issue in many public safety applications, especially those involving the control of reactive components. The latency of mobile communications is high, especially because of the centralization of the core network. Big operators can have a few core network instances deployed throughout a national territory (smaller operators will normally have two), and all the base stations should be connected to one or two of these instances. Therefore, besides the time consumed in the radio access and within the core network, the time used in this interconnection also has to be taken into account. However, in many situations, time is not negligible as it can involve multiple, uncontrolled contractor networks and long distances. The time to transverse the networks is not negligible even if the peers of the communication are camped on the same cell, as all the packets should reach the SGW and the PGW and then return.

In order to improve this situation, two main alternatives are being explored, both of which are based on the provision of Mobile Edge Computing or Fog Computing functionality.

The first approach to reduce latency is the use of an intermediate component called the Fog Gateway (FGW) [GAR 16b], which analyzes the data plane traffic, and is placed between the base station and the serving gateway. The FGW extracts all the relevant information from the encapsulated traffic between the end user and the core network, and when some of the destination addresses are reachable from the FGW domain (e.g. in the case of emergency multicast), it forwards the desired traffic directly, thereby avoiding the transport time spent in the core network. A Message Sequence Chart (MSC) of the procedure is shown in Figure 6.5.

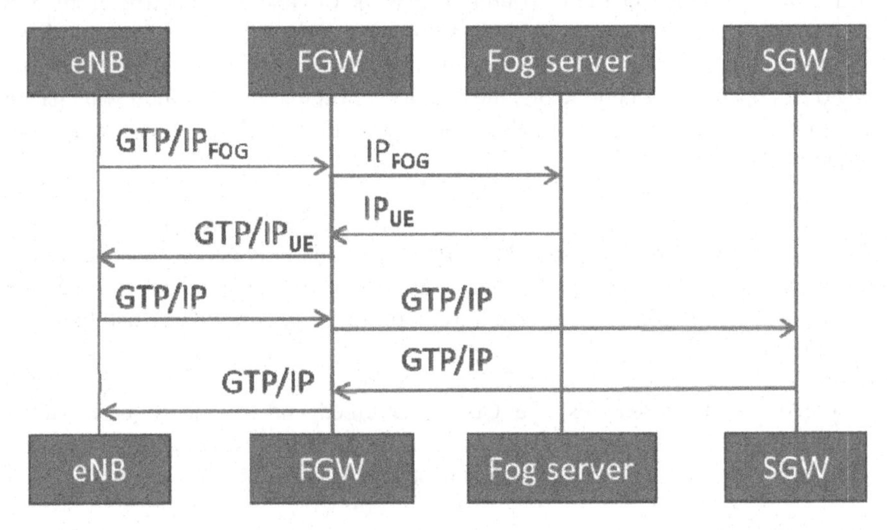

Figure 6.5. *F-GW MSC*

The communication between UEs in the same eNodeB can be greatly accelerated as the FGW will forward these packets in the element immediately behind the cell. In the first experiments conducted in PerformNetworks, latency reduction reached 70% (this value can vary depending on the topology of the backhaul network being considered).

The main disadvantage of this approach is that it only works after receiving the first packet, which may be a problem when the communication is established between two UEs and a handover procedure is triggered, as there may be packet duplication unless a more complex logic is implemented in the FGW.

To overcome this limitation, another approach is also being explored, the deployment of a component, called Middlebox, which provides a similar functionality but does so by analyzing the control plane. The MSC for UE-to-UE communication is shown in Figure 6.6.

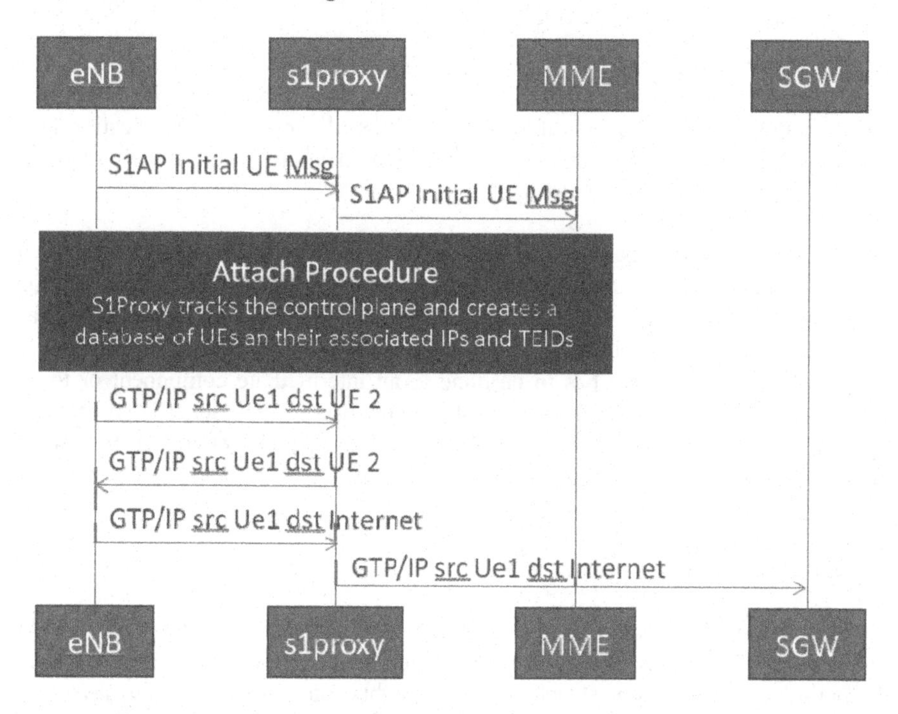

Figure 6.6. *MSC for the Middlebox (named s1proxy in this figure)*

With this new component, a traffic database is established before any data is sent across the encapsulation tunnel so the acceleration can be done from the very beginning of the connection. Furthermore, the component will analyze all the control plane procedures, including the bearer creation and modification procedures or the handover, so the database is always up to date.

The main problem with this approach is security considerations. To guarantee that the intermediate component is able to analyze control plane messages, the ciphering of messages has to be disabled or the component has to have access to the

HSS. The trade-off of analyzing all the messages of the control plane also has to be considered as it will affect all the components on the network.

Both of these approaches can be deployed with SDN techniques. The packet forwarding from the base stations is configured by an SDN controller, and might benefit from SDN switches, which are able to process encapsulated packets up to their transport IP headers, as the components could install rules in the switches to automatically redirect the traffic to an appropriate component.

6.4.3.5. *Group communications*

The Multimedia Broadcast Multicast Services (eMBMS) standard from 3GPP is highly suitable for broadcasting to a large number of users, but for multicast services over a small group of users, it can be an overkill approach. Application-level solutions for implementing this multicast communication suffer latency problems and can congest distribution networks.

The aforementioned components introduced to reduce latency can also be used to implement multicast communications from the network side. To do so, the VELOX engine will install rules in SDN backhaul switches to copy and forward multicast packets (that will be matched with a rule). Depending on the source and destination of packets, the forwarding has to be done to an intermediate component or to the final peer of the communication, i.e. if the destination of the packet is an eNodeB, GTP encapsulation is required, while if it is outside the LTE network of the fog, IP packets can be directly forwarded.

6.5. Conclusions and future work

This chapter has provided an overview of the current possibilities of LTE for public safety services as well as how architecture could be evolved to improve them. To optimize public safety systems, some insights into a methodology have been provided and the major challenges and open questions have also been enumerated.

The number of KPIs currently taken into account by our optimization process will increase in order to characterize the QoE for certain types of traffic such as voice and video. Prototypes for latency reduction components are now being implemented and will be tested at scale with commercial base stations. Several showcases of the full optimization procedure will be performed so as to evaluate the combined impact of all the optimizations.

6.6. Acknowledgments

The work received funding from the European Horizon 2020 Programme under grant agreement 688624 (Q4Health project).

6.7. Bibliography

[AUS 15] AUSTRALIAN GOVERNMENT, PRODUCTIVITY COMMISION, Public Safety Mobile Broadband, Research Report, 2015.

[AZI 15] AZIZ F.M., SHAMMA J.S., STÜBER G.L., "Resilience of LTE networks against smart jamming attacks: Wideband model," *IEEE 26th Annual International Symposium on Personal, Indoor, and Mobile Radio Communications (PIMRC)*, Hong Kong, pp. 1344–1348, 2015.

[BOU 14] BOUET M., PHEMIUS K., LEGUAY J., "Distributed SDN for mission-critical networks," *IEEE Military Communications Conference*, Baltimore, MD, pp. 942–948, 2014.

[CIC 16] CICALÒ S., MAZZOTTI M., MORETTI S. *et al.*, "Multiple video delivery in m-health emergency applications," *IEEE Transactions on Multimedia*, vol. 18, no. 10, pp. 1988–2001, October 2016.

[DIA 14] DIAZ-ZAYAS A., GARCIA-PEREZ C.A., MERINO P., "Third-generation partnership project standards: for delivery of critical communications for railways," *IEEE Vehicular Technology Magazine*, vol. 9, no. 2, pp. 58–68, June 2014.

[DIA 15] DIAZ-ZAYAS A., GARCIA-PEREZ C.A., MERINO P., "PerformLTE: A testbed for LTE testing in the future internet," *Wired/Wireless Internet Communications: 13th International Conference, WWIC 2015*, Malaga, Spain, 25–27 May 2015.

[ETS 10] ETSI, TS 102.900 European Public Warning System, October 2010.

[EUR 11] EUROPEAN NETWORK AND INFORMATION SECURITY AGENCY, "Enabling and managing end-to-end resilience", available at: https://www.enisa.europa.eu/publications/end-to-end-resilience/at_download/fullReport, January 2011.

[FAN 16] FANTACCI R., GEI F., MARABISSI D. *et al.*, "Public safety networks evolution toward broadband: sharing infrastructures and spectrum with commercial systems," *IEEE Communications Magazine*, vol. 54, no. 4, pp. 24–30, April 2016.

[FER 13] FERRUS R., PISZ R., SALLENT O. *et al.*, "Public safety mobile broadband: a techno-economic perspective", *IEEE Vehicular Technology Magazine*, vol. 8, no. 2, pp. 28–36, June 2013.

[FOR 14] FORGE S., HORVITZ R., BLACKMAN C., Is commercial cellular suitable for mission critical broadband? Study on use of commercial mobile networks and equipment for mission-critical high-speed broadband communications in specific sectors, Technical report, European Union, February 2014.

[GAR 16a] GARCIA-PEREZ C.A., RIOS A., MERINO P. *et al.*, “Q4HEALTH: quality of service and prioritisation for emergency services in the LTE RAN stack”, *European Conference on Networks and Communications (EuCNC)*, Athens, pp. 64–68, 2016.

[GAR 16b] GARCIA-PEREZ C.A., MERINO P., “Enabling low latency services in standard LTE networks”, *Foundations and Applications of Self-* Systems (FAS*), 2016 IEEE 1st International Workshops*, pp. 248–255, 2016.

[GER 13] GERMAN FEDERAL MINISTRY OF THE INTERIOR, On the Future Architecture of Mission Critical Mobile Broadband PPDR Networks, White Paper, 2013.

[GUE 14] GUERZONI R., TRIVISONNO R., SOLDANI D., “SDN-based architecture and procedures for 5G networks,” *1st International Conference on 5G for Ubiquitous Connectivity (5GU)*, Akaslompolo, pp. 209–214, 2014.

[LAB 16] LABIB M., MAROJEVIC V., REED J.H. *et al.*, “How to enhance the immunity of LTE systems against RF spoofing,” *International Conference on Computing, Networking and Communications (ICNC)*, Kauai, HI, pp. 1–5, 2016.

[LIC 16] LICHTMAN M., JOVER R.P., LABIB M. *et al.*, “LTE/LTE-A jamming, spoofing, and sniffing: threat assessment and mitigation,” *IEEE Communications Magazine*, vol. 54, no. 4, pp. 54–61, April 2016.

[MOO 16] MOORE L.K., “The First Responder Network (FirstNet) and Next-Generation Communications for Public Safety: Issues for Congress, Congressional Research Service, available at: URL https://www.fas.org/sgp/crs/homesec/R42543.pdf, 2016.

[PEL 15] PELTOLA M.J., HAMMAINEN H., “Economic feasibility of mobile broadband network for public safety and security,” *IEEE 11th International Conference on Wireless and Mobile Computing, Networking and Communications (WiMob)*, Abu Dhabi, pp. 67–74, 2015.

[PRA 16] PRASAD A., MAEDER A., SAMDANIS K. *et al.*, “Enabling group communication for public safety in LTE-advanced networks”, *Journal of Network and Computer Applications*, vol. 62, pp. 41–52, 2016.

[WEE 12] WEERAKKODY V., EL-HADDADEH R., CHOCHLIOUROS I.P. *et al*, “Utilizing a high definition live video platform to facilitate public service delivery”, in *AIAI 2012 International Workshops*, Springer, Berlin-Heidelberg, pp. 290–299, 2012.

[WIR 16] Wireless Emergency Alerts, “Federal Communications Commission”, available at: https://www.fcc.gov/consumers/guides/wireless-emergency-alerts-wea, October 2016.

7

Aerial Platforms for Public Safety Networks and Performance Optimization

Aerial platforms have recently become popular as key enablers for rapid deployable wireless networks where coverage is provided by on-board radio heads. These platforms are capable of delivering essential wireless communications for public safety agencies in the aftermath of natural disasters, and are also able to provide a rapid coverage patch in remote areas which are out of reach of the main public safety network. From this perspective, aerial platforms can be seen as an essential component for a reliable public safety network, since most of the public safety operations are of mission-critical and lifesaving nature. PSN operators can opt to have several standby aerial platforms in order to support emergency off-grid operations or to complement a damaged terrestrial network in case of a natural or man-made disaster. An optimized deployment of these platforms is essential,

Chapter written by Akram AL-HOURANI, Sathyanarayanan CHANDRASEKHARAN, Sithamparanathan KANDEEPAN and Abbas JAMALIPOUR.

considering the fact that over-designing might not be an available choice in disaster circumstances. The optimization process includes the selection of several controllable parameters such as location, transmit power and antenna pattern, but perhaps one of the most important parameters is the altitude of the platform itself.

In this chapter, we provide an overview of aerial platform technology as a key component of public safety networks, discussing the most recent developments in this field including recent projects and initiatives of leading research and industry groups. We also introduce the reader to the nature of the radio propagation properties of the air-to-ground channel that largely differs from ordinary terrestrial radio propagation. Understanding the specific nature of the air-to-ground radio channel will give the necessary background required for the following topic that presents an analytical approach for optimizing the altitude of aerial platforms. This optimization is required to maximize the coverage patch, and to harness the full capabilities of such platforms.

7.1. Aerial supported public safety networks

The main aim of a public safety professional is protecting citizens and property as well as saving lives. As with any other major task, this is not possible without teamwork. As we all know, communication is one of the most important aspects of efficient teamwork which is achieved by the infrastructure provided by public safety networks. Public safety operations rely on a mission critical communication infrastructure for conducting lifesaving and other operational activities, addressing a large number of threats; both natural and man-made, terrorist attacks and accidents. In the aftermath of a disaster, the coordination and effectiveness of the emergency personnel can be dramatically improved by implementing reliable and resilient broadband links connecting different devices, agencies and contacting the headquarters.

One of the challenges faced by the public safety networks is the compatibility between the different technologies used by different public safety organizations such as firefighters, police, medical-aid, etc., ensuring that the coordination activities during the aftermath of a disaster is effective. Moreover, the public safety networks need to cater for today's demand for broadband data, high-definition video, etc. which require higher data rates. Hence, broadband wireless technologies such as the LTE Advanced are seen as a natural successor to traditional technologies used for public safety such as TETRA [BAL 14].

However, public safety networks as any other piece of infrastructure is vulnerable to damages caused by natural and man-made disasters. These disasters can heavily affect network performance or even can fully paralyze it. In this context, reliable disaster recovery solutions are required, which can be swiftly deployed over the afflicted areas. Many different options have been proposed in the literature for public safety networks to enable swift disaster recovery and disaster resiliency features.

One of the proposals includes aerial platforms as a technology enabler for disaster recovery networks [KAN 14, GOM 15, AL 15, CHA 13]. Such platforms are characterized by resilience and robustness, and are capable of being deployed within a few hours after the occurrence of a disaster. In fact, an airborne communication infrastructure concept has been endorsed by the homeland security bureau in the USA. It envisioned the recovery of critical communications for first responders within 12–18 hours [FCC 09]. The ongoing advancements in microelectronics have allowed the recent widespread of small form factor radio heads and processing units, which have a high performance and reduced weight at the same time. Thus, the concept of aerial networks is more feasible and an economic choice in a lot of deployment scenarios where a rapid broadband network is required. Many applications are envisioned for supporting public safety operations in remote areas in the aftermath of natural disasters using aerial platforms. An example of the research and development efforts in airborne network recovery solutions is the European Commission funded FP7 project ABSOLUTE [ABS 13]. This project focuses on Low Altitude Platforms (LAP) to complement a rapidly deployable terrestrial communication infrastructure. Another effort in the field of aerial platforms is being undertaken by Google, providing high-speed Internet access to large areas using high altitude balloons [GOO 14]. Another important application for aerial platforms is foreseen for the commercial cellular industry, where network operators can rapidly deploy aerial platforms to compensate sudden traffic demands during public events, where a massive concentration of users occurs.

Different types of aerial platforms are available to be used for communication purposes. Firstly, they are classified on the basis of their operational height into either high altitude platforms (HAPs) [MOH 11] which operate at 17–22 km in the stratosphere or low altitude platforms which operate at lower heights running up to a few kilometers above the surface of the Earth. Then, they are further classified depending on how they achieve their lift into aerostatic platforms, using either some kind of lifting gas such as hydrogen or helium which makes the platform lighter than air or aerodynamic platforms which use the dynamic forces created by movement through air like an aircraft. Some hybrid aerial platforms which include the advantages from both the types can be found. A special type of aerial platform called Helikite [ALL] was used in the ABSOLUTE project which combined a helium balloon and a kite to form a single tethered aerial platform. Figure 7.1 shows some models of Helikites in operation as well as in a Helibase, ready to be deployed.

Features such as mobility can also be included in some aerial platforms where the aerial platform is either capable of moving with thrusters or uses the winds in the atmosphere or it can be kept quasi-stationary by tethering it to the ground. Other types of aerial platforms might include unmanned aerial vehicles (UAVs), autonomous and remote-controlled quadcopters, etc. which are predominantly utilized for surveillance purposes.

Figure 7.1. *Helikites – a) and b) showing Helikites in operation, c) and d) showing Helikites in a Helibase ready to be deployed (source: Allsopp Helikites Limited [ALL])*

There are many advantages in using aerial platforms. Aerial platforms exhibit additional resilience to disasters such as earthquakes, tsunamis, etc. Aerial platforms can cover large areas compared to terrestrial base stations due to larger favorable line-of-sight conditions for ground terminals with respect to the aerial platform and have lower propagation delays compared to satellites. Aerial platforms can be used as a stand-alone infrastructure providing coverage to a group of ground terminals or can be integrated with an existing terrestrial and/or satellite infrastructure to connect the first responders with headquarters. Figure 7.2 shows one such example of a hybrid integrated aerial–terrestrial network. From the figure, we can see that a terrestrial relay station, where available, can relay the messages from the ground terminals to the aerial platform in addition to the aerial platform providing coverage to other ground terminals [CHA 13]. Also, the aerial platform can communicate with a satellite in order to keep the headquarters informed of the reality on the ground.

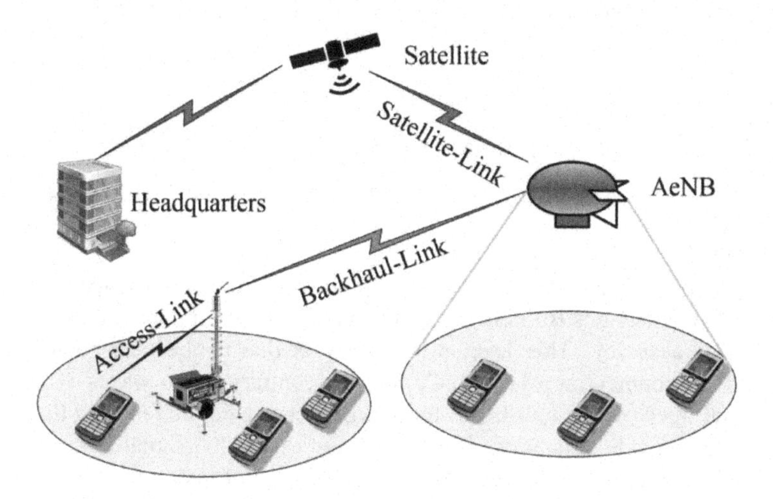

Figure 7.2. *Hybrid integrated aerial–terrestrial network*

7.2. Air-to-ground radio channel

Air-to-Ground (AtG) radio channel has been previously studied in the context of Land Mobile Satellite (LMS), also called Mobile Satellite Systems (MMS), where a mobile station (or a mobile user) could freely roam in an urban environment, rather than in a fixed and highly directional terminal. This capability means heavy radio losses due to the non-line-of-sight nature of the link [CHI 10] especially when compared with legacy immobile VSAT systems. In the case of a LMS system, the service can be provided by a geostationary satellite (e.g. Thuraya), or a constellation of satellites at low/medium Earth orbits (e.g. Iridium, Globalstar) [JAM 97]. Due to the specific nature of the satellite system itself, the previous studies on the LMS radio channel largely depended on the system-specific nature [SCA 08] of the particular satellite under study. The main difference that we will illustrate in the model is that it is system independent rather than environment dependent. In other words, we map the statistical parameters of the urban environment into a radio propagation model, making it more flexible and widely implementable on different urbanized cities and suburbs.

7.2.1. *The nature of the air-to-ground radio channel*

Radio propagation in an AtG radio channel largely differs from legacy terrestrial propagation in cellular networks. For instance, in terrestrial propagation radio models, the mean path loss is usually captured in a log-distance relation, i.e. the path loss is proportional to the logarithm of the distance, having a certain propagation exponent. Namely, a typical terrestrial path loss has the form of [AL 14a]:

$$L = L_{ref} + 10\,\alpha \log d,$$

where α is called the path -> path-loss exponent, L_{ref} is the reference path loss and d is the distance between the transmitter and the receiver, note that the loss here is measured in dB.

The main reasoning behind the wide adoption of the log-distance model in terrestrial communications is firstly due to its mathematical simplicity, and secondly to its level of accuracy. This level of accuracy is due to the near-homogenous propagation environment at a large scale. On the contrary, radio waves in an AtG channel travel freely without obstacles for a large distance before reaching the urban layer of the man-made structures. The latter layer causes the signal to scatter and diffract, leading to an excessive amount of losses on top of the free-space path loss incurred between the aerial platform and the terrestrial user. The nature of the channel between an aerial platform and a terrestrial user is shown in Figure 7.3.

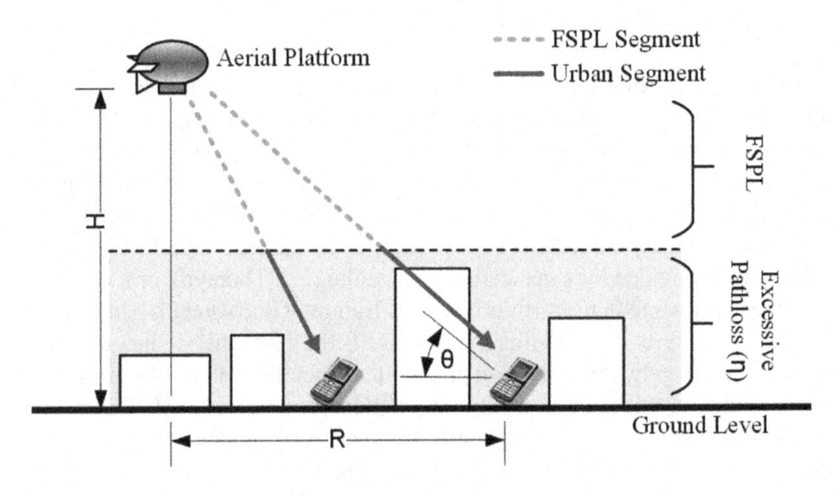

Figure 7.3. *Air-to-Ground propagation between an aerial platform and a terrestrial user*

Accordingly, a better way to model an AtG channel is considering the losses as composed of two parts [AL 14b]: the first part is the free-space path loss (FSPL) and the second part includes the additional losses caused by the urban environment, called the *excessive path loss* and denoted by η. The AtG channel is shown in Figure 7.3, which indicates the two distinct propagation segments, namely the FSPL segment and the urban environment segment. The AtG channel loss can be expressed in dB as follows [AL 14b]:

$$L_\xi = L_o + \eta_\xi, \tag{7.1}$$

where L_o represents the free-space path loss between the aerial platform and the terrestrial user, which is given by:

$$L_o = 20 \log_{10}\left(\frac{4\,\pi\, d f_c}{c}\right),$$

where f_c is the system frequency in Hz, c is the speed of light in meters per second and ξ refers to the *propagation group*. A propagation group represents the set of receivers which share a similar statistical behavior, namely a certain statistical distribution of fading and excessive path loss. It has been empirically noticed that an AtG channel adheres to two main propagation groups that correspond to (1) *good propagation conditions group*, and (2) *bad propagation conditions group*. These distinct performance groups are due to shadowing caused by urban structures, for example they can be referred to as a line-of-sight (LoS) group and a non-line-of-sight (NLoS) group. The probability that a receiver belongs to a certain group depends on the elevation angle θ (i.e. the angle at which the aerial platform is seen from the ground receiver), we call this probability the *group occurrence probability* and we denote it by p_ξ. The excessive path loss is captured in the random variable n_ξ, where its statistics is also dependent on the propagation group.

In the upcoming sections, the reader will learn more about these concepts using practical examples and illustrations.

7.2.2. *ABSOLUTE radio propagation model*

ABSOLUTE is an EU-funded research project [ABS 13], as previously mentioned, it investigated the feasibility of supporting afflicted communication infrastructures in case of a natural disaster, using temporary aerial platforms that are rapidly deployable. One of the important outcomes of this project is the development of a radio propagation model [AL 14b]. This model is mainly used as a solid base for simulations and system performance evaluation.

Radio frequency planners of PSN operators can easily use the ABSOLUTE radio model in order to estimate the expected coverage patch and the performance of aerial platforms. This prediction is simply linked to the statistical parameters of the underlying urban environment describing the city structures as defined by the International Telecommunication Union (ITU-R) in its recommendation document [ITU 03a]. The latter document suggests a standardized model for urban areas based on three simple parameters α_o, β_o and γ_o that describe, to a fair extent, the general geometrical statistics of a certain urban area. These parameters are detailed below:

– parameter α_o represents the ratio of built-up land area to the total land area (dimensionless);

– parameter β_o represents the mean number of buildings per unit area (buildings/km^2);

– parameter γ_o is a scale parameter that describes the buildings' height distribution according to the Rayleigh probability density function:

$$f(h) = \frac{e^{-\frac{h^2}{2\gamma^2}}}{\gamma^2} h .$$

The provided AtG model by ABSOLUTE covers a wide range of possible environments, namely: (1) suburban environment which also includes rural areas; (2) urban environment which is the most common situation and represents average European cities; (3) dense urban environment representing some cities where buildings are in close proximity to each other; (4) high-rise environment representing modern cities with skyscrapers. ITU-R statistical parameters are relatively easy to obtain from a certain city's urban plan, and are likely to be well documented by the city urban planning authorities. Table 7.1 summarizes the selected ITU-R parameters for these environments [HOL 08].

Environment	α_o	β_o	γ_o
Suburban	0.1	750	8
Urban	0.3	500	15
Dense urban	0.5	300	20
High-rise urban	0.5	300	50

Table 7.1. *Selected urban environments*

7.2.2.1. *ABSOLUTE model development*

The ABSOLUTE model is developed using ray-tracing simulations for three types of rays (Direct, Reflected and Diffracted), while transmitted rays are neglected in order to simplify the calculations and because of their very low contribution to the coverage, by transmitted rays we refer to those that penetrate buildings and obstacles. It is important to note that the availability of the optical line-of-sight between the aerial platform and a receiver does not necessarily mean that the RF line-of-sight condition is satisfied, since RF signals require much wider ellipsoids than optical light, i.e. according to Fresnel Zones Concept [SAA 11].

We depict a sample ray tracing for a single receiver in Figure 7.4, where the ray color indicates the intensity of the electric field received by this ray. We can clearly note that the intensity of direct ray (red) is higher than the reflected and diffracted rays, where the resulting electric field is calculated as the vector sum of all captured rays, i.e. by taking into consideration both the magnitude and the phase of the rays.

Figure 7.4. *A simulation snapshot of RF rays between the aerial platform and a single receiver. For a color version of the figure, see www.iste.co.uk/camara/wireless3.zip*

In order to obtain the coverage of the aerial platform over the target area (1,000 × 1,000 m), we simulate the received power of more than 37,000 uniformly distributed receivers, where the electric field of all the captured rays is summed (complex summation) and the received power is calculated from the resulting electric field. The ray-tracing simulator yields a list of all receivers including their corresponding path loss, accordingly we calculate the excessive path loss using [7.1] and obtain its statistical distribution.

For the geometrical structure of the underlying urban environment, we select a model city plan environment similar to Manhattan's grid, as depicted in Figure 7.5, showing an array of structures (buildings or houses) of an assumed square dwelling of width *W* and inter-dwelling spacing *S*.

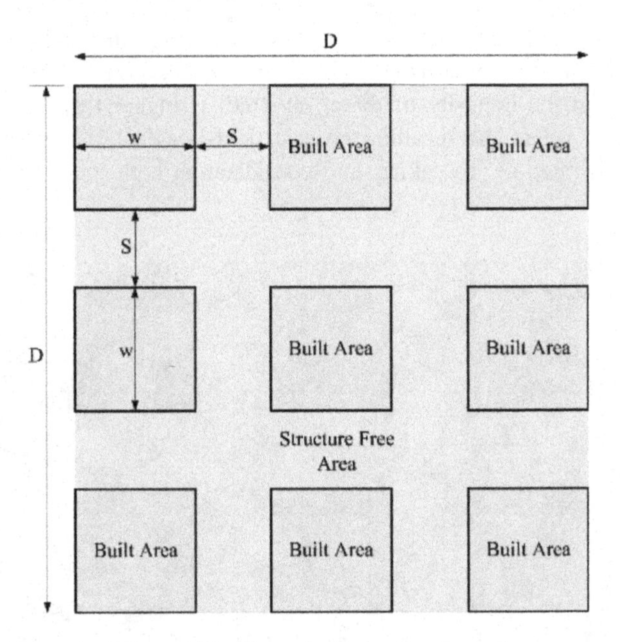

Figure 7.5. *The selected urban layout*

7.2.2.2. *ABSOLUTE model parameters*

We model the excessive path loss as a normal distributed random variable of form $N(\mu_\zeta, \sigma_\zeta)$, where ζ is the propagation group. We use two main propagation groups namely $\zeta = \{G1, G2\}$, where each group has a different statistical behavior.

Starting with μ_ζ, which represents the mean excessive path loss, the simulation did not show a clear dependency on the elevation angle but rather a constant value, obtained by averaging all samples in a certain propagation group. The results of the means are listed in Table 7.2. However, the simulation results suggest that the general trend of the standard deviation of G1 and G2 can be summarized in the following expression:

$$\sigma_{\xi} = a_{\xi}\exp\left(-b_{\xi}\theta\right)$$

where a_{ξ} and b_{ξ} are frequency and environment dependent parameters obtained by curve fitting using the Damped Least-Squares (DLS) method. These parameters are also are listed in Table 7.2.

It is important to note that the ABSOLUTE model is only valid for elevation angles above $15°$, since low elevation angles have a very limited probability of receiving any signal from the aerial platform. Another reason is that the selected simulation setup can produce results for down to about $15°$ only. The resulting empirical curve fitting equation for the group occurrence probability (of group 1) is chosen to balance simplicity and accuracy, and it is a function of the elevation angle, as follows:

$$P_1(\theta) = c(\theta - 15)^d$$

where θ is the elevation angle (the angle at which the AeNB is seen from a certain receiver), c and d are empirical curve fitting parameters, listed in also Table 7.2.

Group 2 occurrence probability is simply the complement of the probability of group 1, and thus it can be calculated as:

$$P_2(\theta) = 1 - P_1(\theta).$$

If the system frequency is not listed in the model parameter (Table 7.2), we can easily obtain the related parameters using interpolation.

We illustrate in Figure 7.6 the difference between using the conventional log-distance path loss model and the aerial path loss model suggested here. The average path loss of the AtG link is calculated as follows (in dB):

$$PL_{ATG} = PL_1 P_1 + PL_2 P_2,$$

where PL_1 is the mean path loss of propagation group (1) that corresponds to receivers favoring line-of-sight condition or near-line-of-sight condition, while the second propagation group (2) generally corresponds to receivers with no line-of-sight but still receiving coverage via strong reflection and diffraction.

	700 MHz			
	Suburban	Urban	Dense urban	High-rise urban
μ_1	0.0	0.6	1.0	1.5
μ_2	18	17	20	29
(a_1, b_1)	(11.53, 0.06)	(10.98, 0.05)	(9.64, 0.04)	(9.16, 0.03)
(a_2, b_2)	(26.53, 0.03)	(23.31, 0.03)	(30.83, 0.04)	(32.13, 0.03)
(c, d)	(0.77, 0.05)	(0.63, 0.09)	(0.37, 0.21)	(0.06, 0.58)

	2,000 MHz			
	Suburban	Urban	Dense urban	High-rise urban
μ_1	0.1	1.0	1.6	2.3
μ_2	21	20	23	34
(a_1, b_1)	(11.25, 0.06)	(10.39, 0.05)	(8.96, 0.04)	(7.37, 0.03)
(a_2, b_2)	(32.17, 0.03)	(29.6, 0.03)	(35.97, 0.04)	(37.08, 0.03)
(c, d)	(0.76, 0.06)	(0.6, 0.11)	(0.36 , 0.21)	(0.05, 0.61)

	5,800 MHz			
	Suburban	Urban	Dense urban	High-rise urban
μ_1	0.2	1.2	1.8	2.5
μ_2	24	23	26	41
(a_1, b_1)	(11.04, 0.06)	(10.67, 0.05)	(9.21, 0.04)	(7.15, 0.03)
(a_2, b_2)	(39.56, 0.04)	(35.85, 0.04)	(40.86, 0.04)	(40.96, 0.03)
(c, d)	(0.75, 0.06)	(0.56, 0.13)	(0.33, 0.23)	(0.05, 0.64)

Table 7.2. *ABSOLUTE model parameters*

We can note from Figure 7.6 that the free-space path loss model cannot be used to represent an AtG channel as it is over-optimistic. We can also compare it with the extended version of the FPSL, which is the log-distance model described earlier, where α is the propagation exponent set as $\alpha = 2.5$ and the system frequency as $fc = 2.6\ GHz$. We can clearly note that the log-distance model cannot represent the AtG channel accurately.

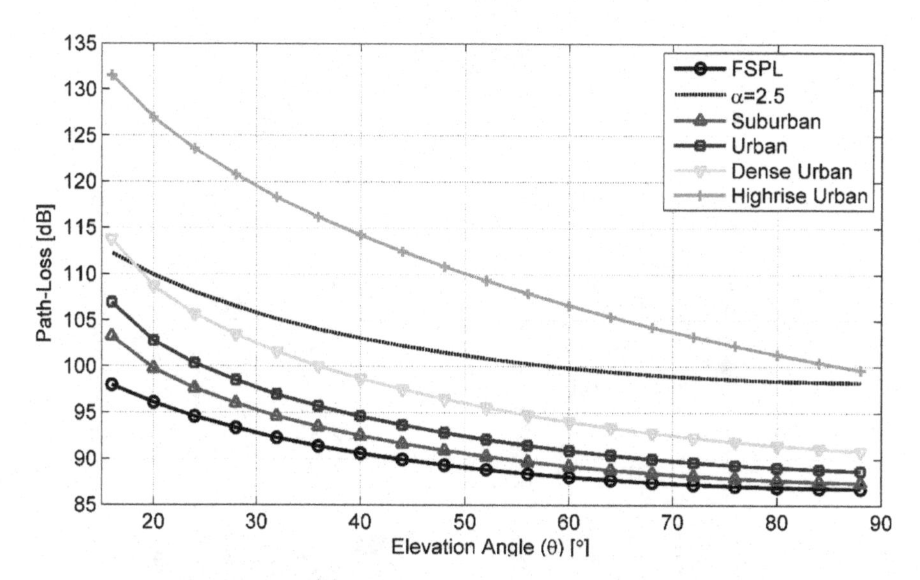

Figure 7.6. *Air-to-ground path loss for different urban environments compared to free-space path loss*

7.2.3. *ABSOLUTE model implementation*

The previous example compared the mean received power in different urban scenarios to the elevation angle. Next we will visualize the effects of shadowing by plotting the heat map of the coverage generated beneath an aerial platform. The received signal level is shown as color variation in Figure 7.7, note that the further we go from the center (where the aerial platform is located above) the less power we receive. The stochastic effect of the shadowing can be observed in figures labeled Suburban, Urban, Dense Urban and High-Rise Urban (see Figure 7.7). For this realization, we utilize a transmit power of +33 dBm at an altitude of 200 m.

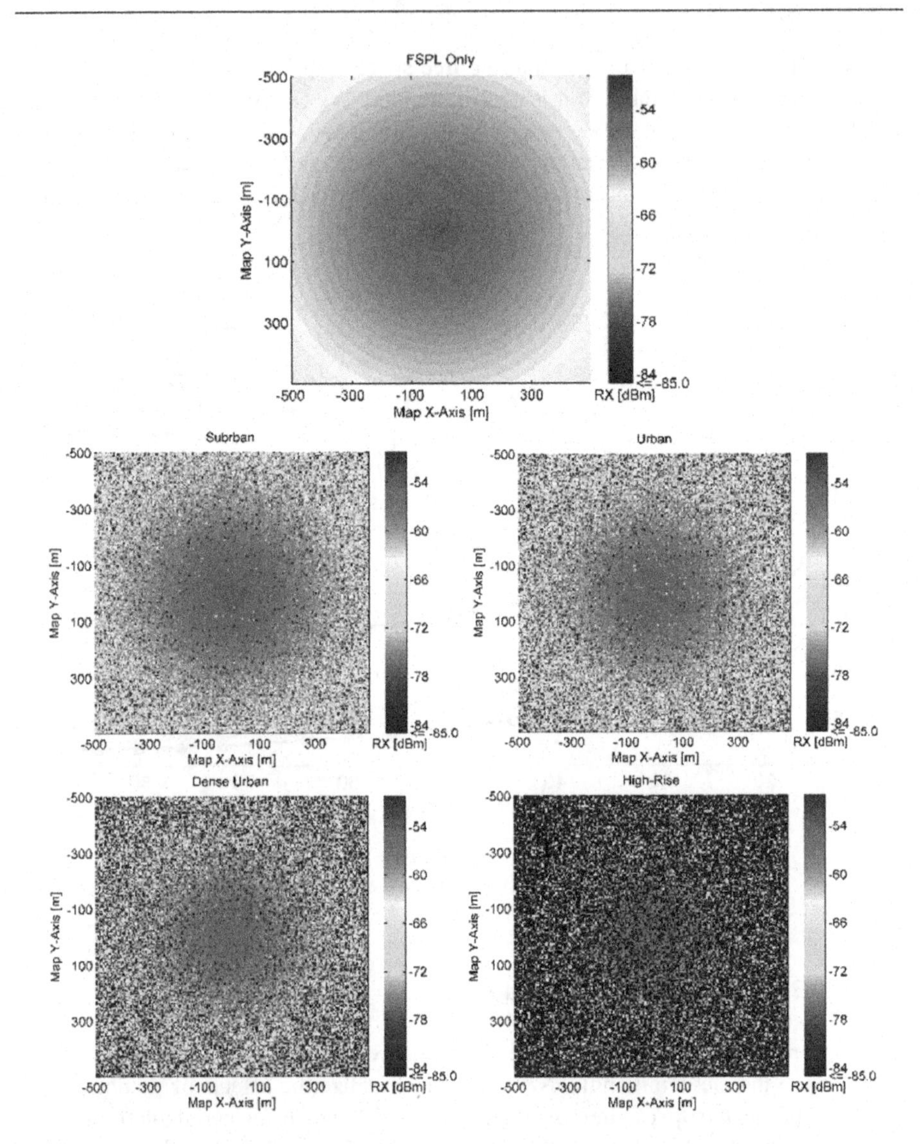

Figure 7.7. *Shadowing variations for different environments. For a color version of the figure, see www.iste.co.uk/camara/wireless3.zip*

In order to simplify the RF design process, we present here the design steps to implement the ABSOLUTE RF model. These steps can be used by RF designers to calculate the expected path loss of ground receivers lying in a certain geographical area. The latter path loss information can then be used to produce information, such as the Signal-to-Interference and Noise Ratio (SINR), and the expected throughput. The steps are presented below:

1) get system parameters (center frequency, altitude) and urban statistical parameters α_o, β_o and γ_o;

2) select a receiver n;

3) calculate the elevation angle θ of receiver n;

4) calculate the free-space path loss of receiver n;

5) pick a propagation group randomly corresponding to the occurrence probability according to a Bernoulli random variable;

6) generate the excessive path loss η of receiver n as a random number, according to the mean and standard deviation of the corresponding propagation group;

7) calculate the total path loss of receiver n;

8) repeat steps 2–7 for all receiving points.

7.3. Optimizing the altitude of aerial platforms

One of the main applications of analytically modeling the AtG channel is the ability to rapidly optimize the performance of an aerial platform, by tuning key design parameters such as altitude. The platform's altitude plays a vital role in determining the size of the coverage area patch, because placing the platform in an arbitrary location might severely degrade its received radio signal on the ground.

It is intuitive to state that higher altitude leads to a lower coverage probability in log-distance path loss models; however, as shown in Figure 7.6, the log-distance does not accurately capture the behavior of an AtG channel and is better represented by the elevation angle. Accordingly, in an AtG channel we observe two contradicting effects when increasing the altitude of a platform; firstly, the probability of getting a line-of-sight increases leading to more users in propagation group 1 (the good group). Secondly, the increasing altitude leads to a higher path loss. Thus, finding the optimum altitude that maximizes the probability of coverage is no longer a trivial approach. In this section, we present a rapid and simple method to effectively optimize the altitude of an aerial platform based on the maximum allowable path loss towards a receiver. We start by modeling the probability of getting line-of-sight in an AtG link.

7.3.1. *Modeling the probability of line-of-sight*

As illustrated previously, the AtG propagation conditions heavily depend on which group a receiver belongs to. We showed that two main propagation groups (receivers groups) are witnessed. The first group corresponds to the good group composed mainly of receivers facilitated with line-of-sight conditions. The second group is the bad group primarily due to the blockage of line-of-sight conditions.

We investigate here the probability of getting a line-of-sight condition based on the ITU document [ITU 03b], where we formulate the resulting LoS probability in a single equation as [AL 14c]:

$$P(LoS)=\prod_{n=0}^{m}\left[1-\exp\left(-\frac{\left[h_{TX}-\frac{\left(n+\frac{1}{2}\right)(h_{TX}-h_{RX})}{m+1}\right]^2}{2\gamma^2}\right)\right],$$

where h_{TX} , h_{RX} are the height of the transmitter and the receiver respectively, and the variable m is given by:

$$m=r\sqrt{\alpha\beta}-1.$$

where r is the ground distance between the transmitter and the receiver, as shown in Figure 7.3, and $\lfloor . \rfloor$ is the floor operator (nearest integer that is smaller than the value of the operand). It is worthy to mention that the geometrical LoS is independent of the system frequency.

In the particular case of an aerial platform we can disregard h_{RX} since it is much lower than the average buildings' heights and the aerial platform's altitude. Also, the ground distance becomes $r=h\tan(\theta)$, where h is the altitude. It is important to note that the resulting plot of the above equation will be smooth for large values of h, accordingly the line-of-sight probability can be considered as a continuous function of θ and the environment parameters. Plotting this probability in Figure 7.8 for four selected urban environments: Suburban (0.1, 750, 8), Urban (0.3, 500, 15), Dense Urban (0.5, 300, 20) and High-rise Urban (0.5, 300, 50) for α ,

β and γ respectively, we note that the trend can be closely approximated to a simple modified sigmoid function (S-curve) of the following form:

$$P_{LoS}(\theta) = \frac{1}{1 + a\exp(-b[\theta - a])},$$

where a and b are called the S-curve parameters. This approximation significantly eases the calculation of the LoS probability, and also allows analytical tractability. In order to generalize the solution, we have linked the S-curve parameters a and b directly to the environment variables α, β and γ, as explained in detail in [AL 14c].

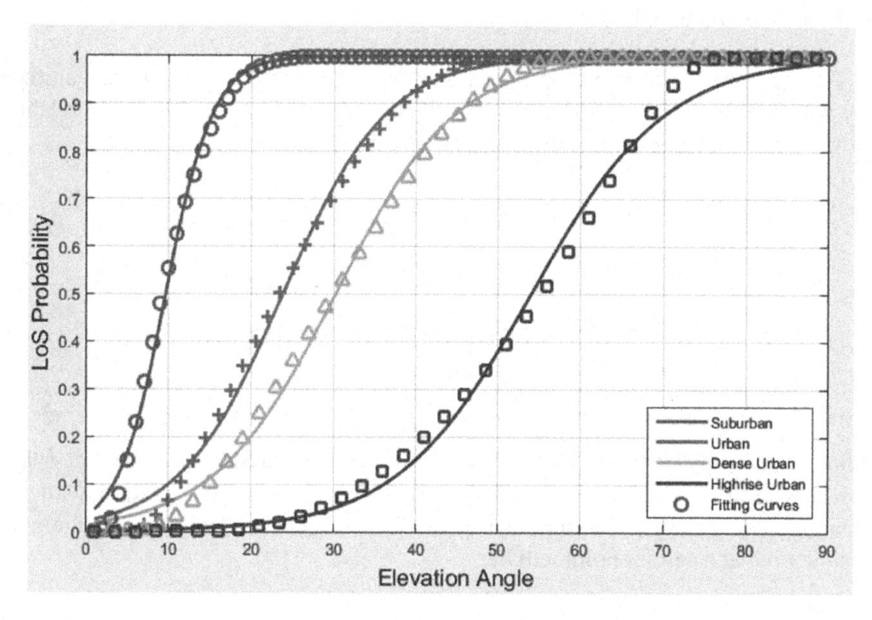

Figure 7.8. *Line-of-sight probability in different urban environments versus the elevation angle*

7.3.2. *Modeling and optimizing system performance*

The performance of a communication system largely depends on the goodness of the channel between the sender and the receiver, namely it depends on the Signal-to-Noise and Ratio (SNR), along with other parameters such as delay. Many performance metrics need to be taken into consideration to satisfy a certain quality of service (QoS) requirement; however, two of the important metrics are (1) the throughput (data rate) between the aerial platform and the ground receiver; (2) the link success probability, which can be seen as the ratio of the number of receivers satisfying the required QoS to the total number of receivers. These two metrics directly depend on the SNR. Thus we can say that a certain QoS requirement directly mandates the target SNR, which eventually mandates the maximum allowable path loss between the aerial platform and terrestrial users, and assuming that both transmit and noise powers are constant, we denote this quantity as PL_{max}.

When the total path loss between the aerial platform and a receiver exceeds this threshold, the link is deemed as failed. For ground receivers, this threshold translates into a coverage disk (zone) of radius *R*, since all receivers within this disk have a path loss that is less than or equal to PL_{max}. Figure 7.9 shows the concept of the coverage region of an aerial platform, where users located at the edge of the region incur the maximum path loss.

Accordingly, we form our optimization problem which is to find the best altitude that will maximize the coverage of radius R. In order to do so, we deduce a relation between the altitude h and the cell radius R as follows:

$$\mathrm{PL_{LoS}} = 20\log d + 20\log f + 20\log\left(\frac{4\pi}{c}\right) + \eta_{\mathrm{LoS}}$$

$$\mathrm{PL_{NLoS}} = \underbrace{20\log d + 20\log f + 20\log\left(\frac{4\pi}{c}\right)}_{\mathrm{FSPL}} + \underbrace{\eta_{\mathrm{NLoS}}}_{\eta_{\xi}}$$

where d is the distance between the aerial platform and a receiver in a circle of radius r, given by $d = \sqrt{h^2 + r^2}$, while f is the system frequency. The free-space path loss (FSPL) is given according to the Friis transmission equation with the assumption of isotropic transmitter and receiver antennas. Thus, the average path loss incurred at a certain point will be:

$$PL_{ATG} = P_{LoS}(\theta) \times PL_{LoS} + P_{NLoS}(\theta) \times PL_{NLoS}$$

We can therefore write:

$$PL_{max} = \frac{A}{1 + a\exp\left(-b\left[\arctan\left(\frac{h}{R}\right) - a\right]\right)} + 10\log\left(h^2 + R^2\right) + B,$$

where the values of A and B are given by:

$$A = \eta_{LoS} - \eta_{NLoS}$$

$$B = 20\log f + 20\log\left(\frac{4\pi}{c}\right) + \eta_{NLoS}$$

The formula describing PL_{max} is implicit, where neither R nor h can be written as an explicit function of each other. In order to obtain the optimum point of the aerial platform altitude h_{OPT} that yields the best coverage, we need to search for the value of h that satisfies the equation of the critical point:

$$\frac{\partial R}{\partial h} = 0,$$

i.e. the point at which the radius–altitude curve in the PL_{max} equation changes its direction. The optimum altitude of an aerial platform is strongly dependent on the specific urban environment condition, in order to illustrate this dependency we have plotted different radius–altitude curves according to the PL_{max} equation. These curves result from the four selected urban environments while maintaining a constant $PL_{max} = 110$ dB, the utilized system frequency is $f = 2{,}000$ MHz, and the excessive urban path loss (η_{LOS}, η_{NLOS}) pairs (0.1, 21), (1.0, 20), (1.6, 23), (2.3, 34) corresponding to Suburban, Urban, Dense Urban, and High-rise Urban respectively (measured in dB).

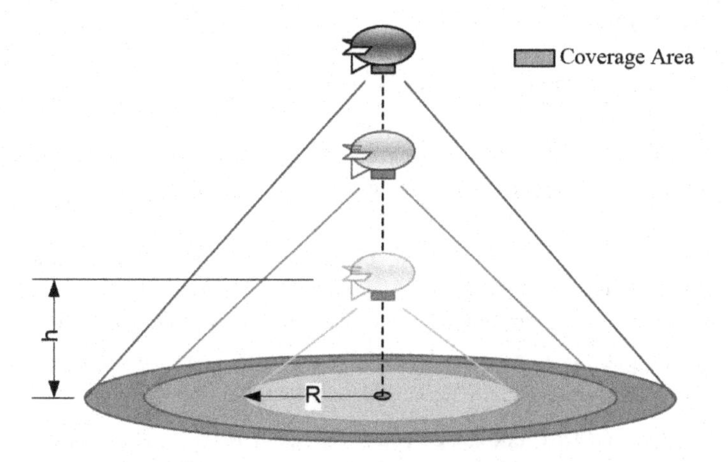

Figure 7.9. *The coverage zone of an aerial platform. For a color version of the figure, see www.iste.co.uk/camara/wireless3.zip*

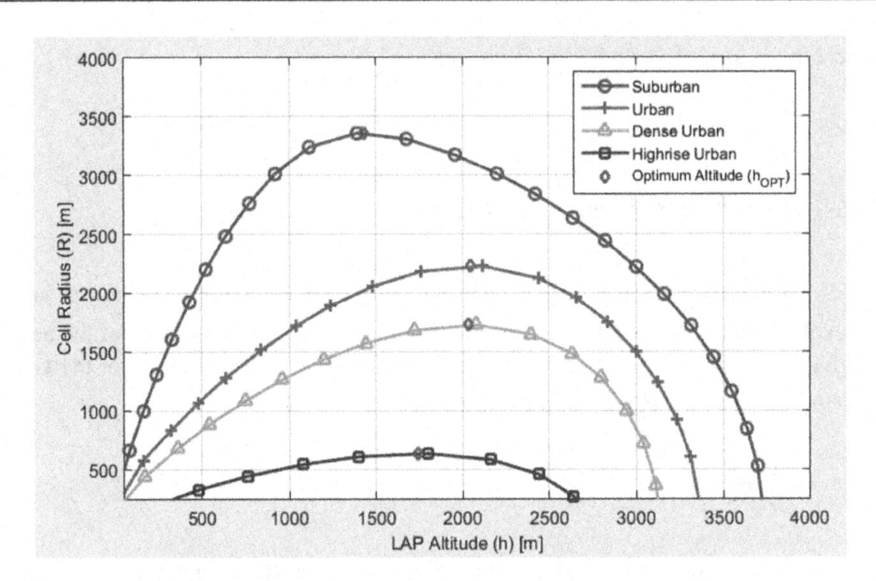

Figure 7.10. *Aerial platform altitude optimization in order to maximize the coverage region for certain* PL_{max}

Here we note that if the maximum permitted path loss is sufficiently high, then the theoretical optimum altitude might reach a very high value for practical implementation. In this case, the aerial platform should be placed at the maximum altitude allowed by the mechanical properties of the platform.

7.4. Bibliography

[ABS 13] ABSOLUTE, Absolute EU project, available at: http://www.absolute-project.eu/, 2013.

[AL 14a] AL-HOURANI A., CHANDRASEKHARAN S., BALDINI G. *et al.*, "Propagation measurements in 5.8 GHz and pathloss study for CEN-DSRC", *International Conference on Connected Vehicles and Expo ICCVE*, 2014.

[AL 14b] AL-HOURANI A., KANDEEPAN S., JAMALIPOUR A., "Modeling air-to-ground path loss for low altitude platforms in urban environments", *Globecom 2014, Symposium on Selected Areas in Communications: Satellite & Space Communication*, 2014.

[AL 14c] AL-HOURANI A., KANDEEPAN S., LARDNER S., "Optimal LAP altitude for maximum coverage", *IEEE Letters on Wireless Communications*, vol. 3, no. 6, pp. 569–572, 2014.

[AL 15] AL-HOURANI A., CHANDRASEKHARAN S., JAMALIPOUR A. *et al.*, "Optimal cluster head spacing for energy-efficient communication in aerial-backhauled networks", *IEEE Globecom*, pp. 1–6, San Diego, 2015.

[ALL 14] ALLSOPP HELIKITES LIMITED, available at: http://www.allsopp.co.uk/, 2014.

[BAL 14] BALDINI G., KARANASIOS S., ALLEN D. *et al.*, "Survey of wireless communication technologies for public safety", *IEEE Communications Surveys & Tutorials*, vol. 16, no. 2, pp. 619–641, 2014.

[CHA 13] CHANDRASEKHARAN S., KANDEEPAN S., EVANS R. *et al.*, "Clustering approach for aerial base-station access with terrestrial cooperation", *IEEE Globecom Workshops*, Atlanta, pp. 1397–1402, 2013.

[CHI 10] CHINI P., GIAMBENE G., KOTA S., "A survey on mobile satellite systems", *International Journal of Satellite Communications*, vol. 28, no. 1, pp. 28–57, 2010.

[FCC 09] FCC, The role of deployable aerial communications architecture in emergency communications and recommended next steps, Report, 2009.

[GOM 15] GOMEZ K., HOURANI A., GORATTI L. *et al.*, "Capacity evaluation of aerial LTE base-stations for public safety communications", *EuCNC 2015*, pp. 133–138, Paris, 2015.

[GOO 14] GOOGLE, available at: http://www.google.com/loon/, 2014.

[HOL 08] HOLIS J., PECHAC P., "Elevation dependent shadowing model for mobile communications via high altitude platforms in built-up areas", *IEEE Transactions on Antennas and Propagation*, vol. 56, no. 4, pp. 1078–1084, 2008.

[ITU 03a] ITU-R, Propagation data and prediction methods required for the design of terrestrial broadband millimetric radio access systems operating in a frequency range of about 20–50 GHz, P-Series, Radioware propagation, Report, 2003.

[ITU 03b] ITU-R, "Propagation data and prediction methods for the design of terrestrial broadband millimetric radio access systems", Rec. P.1410-2, P-Series, Radiowave propagation, Report, 2003.

[JAM 97] JAMALIPOUR A., *Low Earth Orbital Satellites for Personal Communication Networks*, Artech House, Inc., Norwood, 1997.

[KAN 14] KANDEEPAN S., GOMEZ K., REYNAUD L. *et al.*, "Aerial-terrestrial communications: terrestrial cooperation and energy-efficient transmissions to aerial base stations", *IEEE Transactions on Aerospace and Electronic Systems*, vol. 50, no. 4, pp. 2715–2735, 2014.

[MOH 11] MOHAMMED A., MEHMOOD A., PAVLIDOU F.N. *et al.*, "The role of high-altitude platforms (HAPs) in the global wireless connectivity", *Proceedings of IEEE*, vol. 99, no. 11, pp. 1939–1953, 2011.

[SAA 11] SAAKIAN A., *Radio Wave Propagation Fundamentals*, Artech House, 2011.

[SCA 08] SCALISE S., ERNST H., HARLES G., "Measurement and modeling of the land mobile satellite channel at Ku-band", *IEEE Transactions on Vehicular Technology*, vol. 57, no. 2, pp. 693–703, 2008.

8

Topology Control for Drone Networks

8.1. Introduction

The first hours in disaster scenarios are crucial, because it is when rescuers have the highest probability of finding people alive, but unfortunately it is also when resources are scarce. In many disaster scenarios, such as earthquakes, avalanches, mudslides and explosions, it is common for people to be injured and covered by debris. These people require rapid attention and all available human resources need to concentrate on the search and rescue task. Nevertheless, the rescue efforts need to be coordinated and the disaster's impact need to be characterized as fast as possible. The time pressure, scarcity of human resources and urgent need for information are more than enough reasons that justify the use of autonomous drones to collect the maximum possible information and provide communication over the affected areas.

Chapter written by Daniel CÂMARA.

American Red Cross issued a report where it states that aerial drones are one of the most promising and powerful new technologies to improve disaster response and relief operations [AME 15]. The report also confirms that drones naturally complement traditional manned relief operations by helping to ensure that operations can be conducted safer, faster and in a more efficient way. This shows the importance and the interest of the use of autonomous robotic elements in the public safety field. Robotic elements in their various forms, unmanned aerial terrestrial and maritime vehicles can be considered as tactical systems that can help emergency professionals in their missions by providing information in real time and being present in disaster sites, even if the conditions are too dangerous for humans or other living beings such as search and rescue dogs.

This work focuses on drones or unmanned aerial vehicles (UAVs), because when a disaster strikes, it is vital to assess the extent of the damage and the accessibility to the affected areas thus providing an up-to-date assessment of the present status of the disaster site. Traditionally, rescue operation teams acquire such information from satellite imagery analysis [MAD 15] and rescue teams also rely on satellite communication in disaster areas. Satellite imaginary/communication are mature and reliable technologies; they have a series of advantages, notably their wide area coverage and the ability to capture multimodal information such as ultraviolet, visible and infrared spectra. Satellites are also unaffected by ground weather conditions. These characteristics make satellites ideal solutions for long-term planning in disaster areas and as a tool to provide a broad view of the present situation. However, access to the latest satellite data can be costly. For this reason access to satellite data, in case of a disaster, is governed by international agreements such as the International Charter on Space and Major Disasters [INT 16]. Such agreements ensure the area affected by a disaster can have a unified system for accessing space-based data over the affected region. On the other hand, satellite-based image acquisition suffers from a couple of disadvantages. The first one is the relatively low image resolution. After that, you have the variation on the revisit time, which can vary from a couple of hours to the range of weeks. Finally, the images are taken with a high looking downward viewpoint, what makes its use for performing damage assessment on the affected areas extremely difficult. On top of that, factors such as shadowing and low visibility can significantly decrease the quality of satellite imagery and make assessment even harder. In terms of communication, satellites may present delays and uplink/downlink bandwidths that can considerably impact the applications using it. Drones can help to bypass all of these satellite limitations, first they are easy to deploy, easy to handle and relatively affordable (prices vary from \$500 to \$20,000). They also have a high degree of maneuverability, which means they can take pictures and capture data from a much wider range of viewpoints than satellites [UNI 16] with fast high-resolution views from different angles of the affected area, and can be used to provide broadband network access, with small delays, to rescuers and citizens in disaster areas.

UAVs are already being used with success in disaster relief efforts, for example in the recent Nepal earthquake, there were at least seven humanitarian UAV teams operating [KWO 15]. In a recent survey [SOE 16] from UAViators [HUM 16] and Swiss Foundation for Mine Action (FSD) [SWI 97] shows that 70% of people in the humanitarian aid field see the use of drones either as a positive or as a neutral way. The reasons cited for a negative perception in general linked to concerns that the technology creates a distance between beneficiaries and aid workers; the potential association with military applications; and the lack of added value delivered by the use of drones [SOE 16]. Of the 51 respondents who used drones or looked into their use, only three viewed them unfavorably. Among the uses that people associate with drones are mapping, monitoring, search and rescue delivery and public information. Another advantage of drones is that they may perform a series of different tasks at the same time. For example, not only they can act as access points for rescuers devices, but also, at the same time, they can map the region and even make an infrared scan of it.

All of these applications need the establishment of a communication network between a number of different computational elements, such as unmanned vehicles, access points, rescuers' portable devices or the central command office. In fact, communications have become a key concern in large crisis events network, primarily when it involves numerous organizations, human responders and an increasing amount of unmanned systems which offer precious but bandwidth hungry situational awareness capabilities [ICA 16].

This chapter will concentrate on the establishment, and maintenance, of an on-demand topology for unmanned mobile computational devices. To accomplish their tasks, drones need to act as fast as possible and in a self-organizing way. The main objective is to grant a stable communication network among these equipment so that they can efficiently perform their tasks. The less time we take to evaluate the damages and identify victims the greater the chances of saving lives. To increase the efficiency of drones, and decrease the time required to cover the target region, we can multiply the number of drones and, autonomously, parallelize the work to be done. Even though drones are capable of providing a large amount of information, it is unlikely that people in the terrain will have the time and human resources available to control drones and guarantee the full perfect coverage of a given area. Much less if we are talking about a fleet of distinct and purpose specific drones, which would require different and specialized skills. This work tackles exactly this subject, the required mechanisms to provide a useful and autonomous drone fleet.

The remainder of this chapter is organized as follows. In section 8.3, we introduce the related work. In section 8.4, we discuss some possible applications of drones over disaster scenarios. In section 8.5, we present the internal mechanisms that all drones should implement. In section 8.6, we present the fleet architecture and the different tasks that each one of the drones is more prone to execute. Finally, in section 8.11, we conclude the chapter.

8.2. Scenario

Disaster scenarios are, in general, chaotic places, mainly for the first wave of relief effort teams. Access may be difficult and the resources for the rescue operation teams relatively limited. As previously stated, the first hours after a disaster are determinant but are also the more turbulent ones. Communication capabilities in the target region may be lost, the very map of the region may be outdated, the access to areas may have been compromised, e.g. debris, flooded roads, collapsed buildings. This combination of factors, plus the scarcity of human resources and the feeling of urgency in finding/contacting affected people demands an efficient and organized set of measures. Some of these, typically, require specialized rescue professionals, but automated drones could perform others. This would free the rescuers from their burden, and allow them to focus their attention on fundamental tasks.

The task drones can excel at performing are, normally, non-intrusive ones, and can go from, for example, providing network access over the affected area and automatically mapping the changes on the region topography to the detection of victims using a series of distinct sensors. Figure 8.1 shows a possible topology for drones, where different drones perform different tasks over a defined area. We should ensure that we should always try to use the right tool to perform the task at hand; different tasks may require different drones. However, one thing is general; over disaster scenarios, drones should always be as autonomous and independent as possible, so that they are operational without draining the scarce human resources. By autonomous we mean that drones should have their own missions to perform and they complete each mission, or set of missions, without any human interference. Ideally, they would receive their mission, or a set of missions and priorities, from the command operation center, and work on each one until it is fulfilled, and then they demand a new set of missions. As far as possible the drones should be autonomous – even tasks such as recharging should be automatic – so that they perform without human interference. This is an important point because, regardless of the powering model used, e.g. electric or internal combustion engines the available energy is always limited and, at some point, it will end. Moreover, drones may work continuously on their missions 24/7, fulfilling mission after mission. Human operators, on the other hand, cannot, so it is reasonable to assume automatic mechanisms to optimize drones availability.

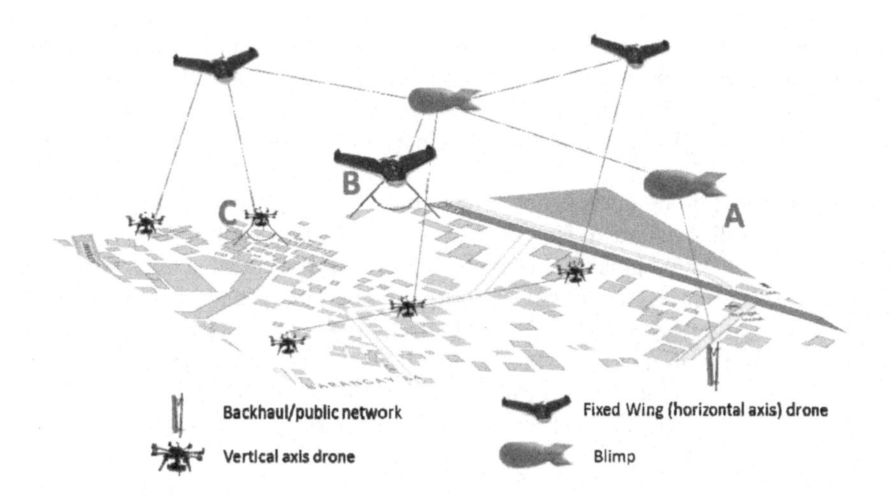

Figure 8.1. *Example of deployment of a mission-based airborne network, where different nodes perform different tasks providing different kinds of information and services. Node A is a blimp squadron, which is stable and has a large autonomy, and can be used as a network backbone to provide connection to the other nodes. Node B is a fixed wing node that can be used to take aerial photos to automatically determine the up-to-date cartography of the affected region. Node C is a vertical axis drone, which can hover over a specific point trying to sense signals and pinpoint possible survivors*

8.3. Related work

There are many use cases where drones have already being useful in humanitarian settings. For example, the United Nations Office for the Coordination of Humanitarian Affairs (OCHA) released a study on the use of UAVs in humanitarian responses [GIL 14]. This report presents three different use cases of real settings where UAVs were successfully used in different disaster scenarios. The first is the damage assessment in the Philippines after Typhoon Haiyan, where drones were used to identify the best place to set up a base of operations and then to check which roads were blocked. The second use case is mapping and disaster risk reduction in Haiti, it discusses how drones were used in the aftermath of the 2012 hurricane Sandi in Haiti. The third use case is Peacekeeping in the Democratic Republic of Congo, where drones were used for surveillance and protection of refugee camps.

Danoffice IT has a commercial drone solution for disaster response [PED 14]. It has already been used in real operation sites such as the typhoon Yolanda at Tacloban, Philippines, where it helped with the planning of the operations site, and on the identification of which roads were usable after the typhoon. In the same disaster, the CorePhil DSI team [MEI 14] used a fixed wing drone, eBee, to capture aerial imagery of downtown Tacloban. These images were further analyzed through crowdsourcing and helped with the generation of the most detailed and up-to-date maps of the region. These maps were used by different humanitarian organizations and even by the Filipino Government for the planning of the relief efforts.

The control of drone fleets is not really a new theme. In fact, it is a well-studied subject in the military context. However, even there the fleet control proposals focus mainly on how to help humans control the fleet, not on having an autonomous fleet. For example, Cummings *et al.* [CUM 07] proposed an automation architecture to help humans on the supervision of a drone fleet. However, the drones are not fully autonomous, a human operator who decides the actions is always in control of the mission. The same comments are valid for other works in the field, e.g. the work of Arslan and Inalhan [ARS 09], where the whole effort relies on helping one operator to control multiple drones.

8.4. Examples of drone applications

In a disaster scenario, drones can perform a number of different tasks to help with the relief effort. Tasks may vary from providing communication to the creation of high-resolution maps of the area and the autonomous search for victims.

Maintaining communication over disaster areas is challenging. We cannot just rely on the public communication network, firstly, because it may be unavailable in remote areas and, secondly because even if it is available the network may be damaged or destroyed. Nevertheless, the coordination of the relief efforts requires communication. Drones can work as temporary mobile access points for extending the coverage on affected areas. This service may be offered not only for the rescuers, but also for the general population with the creation of small picocells. For example, after hurricane Katrina, at New Orleans, the public network was out of service and Verizon, the local provider, gave the first responders the right to use their frequencies.

Another important task, that can be autonomously performed, is the creation of high-resolution maps of the affected area. Disasters may drastically change the affected region, which may void previous maps completely. Drones can fly over the region with 3D cameras and, with the help of GPSs and publicly available relief maps of the region, automatically create up-to-date 3D maps of the area. These maps can be used to understand the impact of the disaster over the region and, for example, decide which roads need to be closed, find the best paths to reach the most damaged areas or even plan the delivery of relief supplies.

As for active roles, they can play to help on the search and rescue operations, we can highlight the infrared scan of the region, trying to find people on/under the debris. Ground penetration radars may be used to find people buried under the debris, and the signal from portable scanning devices can also be used for the same purpose.

Airborne networks are wireless communication networks, which are formed by nodes such as satellites, balloons, drones or fixed wing aircrafts. Figure 8.2 shows some of the technologies that can be involved. These kind of networks have their origins in military environments, flying nodes are used to create an access network over hostile environments. However, recently we have seen the development of civil interest in drone networks. For example, Google's Loon [DEV 13] wants to provide Internet access to populations without access. The project seeks to create a hierarchical airborne network of high altitude balloons, traveling on the edge of space around 20 km above the Earth's surface. At this altitude, the balloons will move with the wind currents in the upper atmosphere to provide access to connect people in rural and remote areas and help fill coverage gaps, and bring people back online after disasters [GOO 15]. The intention is to establish partnerships with local telecommunications companies to share the cellular spectrum and enable connection to the balloon network directly with LTE-enabled devices. The signal is then routed over the balloon network and, at some point, back down to the global Internet on Earth. Each balloon can provide connectivity to a ground area of about 40 km in diameter.

Facebook has also demonstrated interest in flying technologies to provide Internet access to uncovered areas. They recently acquired Ascenta, a UK-based company that designs solar-powered drones, and Mark Zuckerberg, the head of Facebook, has written a post stating they are "working on ways to beam Internet to people from the sky" [GAR 14]. In a press release [ZUC 14], they discuss some of the different technologies that can be used to provide aerial solutions for connectivity, and that different populations have different needs. However, taking into account the number of restrictions and challenges for a project of such enormous size, a network of drones flying at an altitude of 65,000 feet (~20 km) is viewed as an ideal access-providing mechanism. Even though few of the techniques used to interconnect the drones, and the different technologies, have been released, among the advantages for a drone network stated in the press release are [FAC 14]:

– with the efficiency and endurance of high altitude drones, it is possible that an aircraft could remain in the air for months or years, much longer periods than their balloon counterparts;

– unlike balloons, which drift on the wind with limited controls, drones can remain over a specific city or area;

– in case of the need for maintenance, drones can easily be returned to the Earth and redeployed.

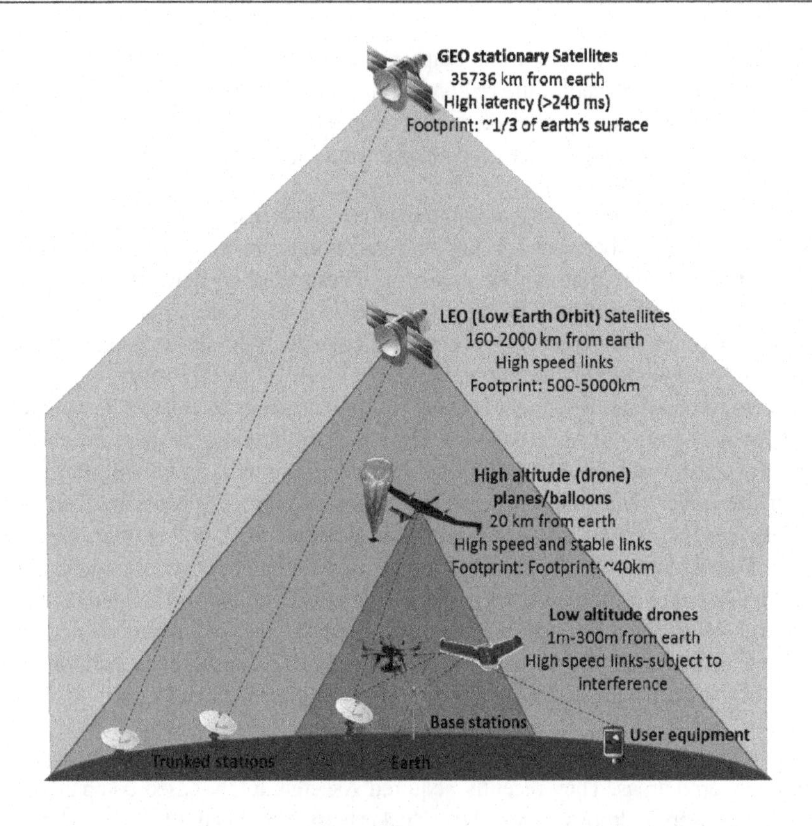

Figure 8.2. *Main available aerial communication methods used in disaster scenarios*

The topology management and control of airborne networks is a critical subject for the network's success, given the dynamicity of the network. The links among the nodes must be easily and quickly reconfigurable to respond to the mobility of the flying nodes. Aerial nodes can change flight paths and missions often; moreover, sudden changes in climatic conditions may have great influence on the nodes mobility. Automatic and dynamic topology management mechanisms play a fundamental role in maintaining the network connectivity. Although still a young research field, different organizations, with different objectives, were proposed on the context of airborne networks. Milner *et al.* [MIL 03] propose the formation of a backbone network with Airborne Networking Platforms (ANPs) to enhance the reliability and scalability of tactical networks. This set of ANPs may be seen as mobile base stations with predictable and well-structured flight paths and the possible combat aircrafts on a mission are the mobile clients. The topology control mechanism proposed here can be used to maintain such topology, but it goes even further in the sense that it adapts automatically to different kinds of missions. By controlling a small set of configuration parameters we can not only maintain a backbone, but also change the way in which the other nodes are interconnected, taking into account the mission at hand. We are interested in maintaining a stable connected network for the backbone, but also we want our topology management mechanism to handle the connectivity of the drones in mission, either in a specific formation or not.

A relatively simple way to improve network connectivity is to have large transmission ranges for the ANPs. However, the large transmission range also implies high power consumption. In order to decrease the power consumption, targeting the improvement of the network lifetime and the autonomy of the drones, we can control the power of the transmission to decrease the range of the nodes. The smaller the range, the smaller the interference among users. However, in this case the role of the topology management mechanism in maintaining the connectivity becomes even more important. Shirazipourazad, Ghosh and Sen [SHI 11] define a critical transmission range (CTR) to be the minimum transmission range of the ANPs to ensure that the dynamic network formed by the movement of the ANPs remains connected at all times, even in presence of attacks and nodes failures. However, they target mainly one specific scenario: providing access to other aerial nodes. They also consider that the full network flight plan is known in advance and does not change greatly during the mission, what is a strong assumption, mainly for drone networks in disaster areas.

Krishnamurthi *et al.* [KRI 09] present MAToC (Mission Aware Topology Control), a topology control mechanism for the backbone of airborne networks. MAToC uses a distributed protocol to exchange the nodes flight plan and uses these collected plans to assign optimal power, channel and bore-sight direction to the airborne antennas. MAToC uses a geometric optimization methodology to assign antenna powers to maximize Signal-to-Interference and Noise Ratio (SINR). It also constantly monitors the links in order to search for broken links to fix later; the feedback comes either through the routing layer or through proactive Hello/Hello-Ack messages. To ensure the airborne network connectivity, backup links are maintained to replace possible faulty links. Link failure can occur in highly dynamic airborne networks because of factors such as mobility, interference or jamming. A well-known solution is to maintain a secondary communication channel that can be set up when the original link fails [GAR 14]. One of the main ideas behind MAToC is that in networks which utilize omni-directional antennas, typical for networks composed of tactical aircraft, the only means of controlling the topology is by varying the transmit power. While keeping a high transmit power may improve connectivity and reduce the number of hops required through a network, it also increases interference and complexity of routing. However, Krishnamurthi *et al.* do not consider that other solutions, like different frequencies and dual radios could also be considered.

Midkiff and Bostian [MID 02] present a two-layer network deployment method to organize PSNs. Their network consists of a hub, and many possible purpose specific routers, to provide access to nodes in the field. In some sense our work provides the same kind of topology, since we are interested in the backbone creation to provide access for the end nodes, e.g. firefighters in the field. However, Midkiff and Bostian's work has two characteristics that we want to avoid. First, the hub represents a single point of failure. If something happens to it, all the communication would be down, even between nodes inside the field. It is important for public safety networks to be as resilient as possible. The second issue we want to avoid, if possible, is long-range communications and the fact that all transmissions pass through the hub. One of the objectives of this work is to avoid, as much as possible, single points of failure, while ensuring the availability of local communications. Narrowing communications to the areas they are really needed, we save resources for other transmissions that really need to cross the network.

The EU FP7 ABSOLUTE [FP7 12] investigates the possibility of deploying hybrid aerial-terrestrial network architecture, in order to provide LTE broadband communications over a large disaster area (Figure 8.3). The access would be granted using specific purpose Aerial eNB (AeNB) and a Low Altitude Platform (LAP) to provide aerial coverage to a large area, and portable lightweight Terrestrial eNB (TeNB) to enhance network capacity in hotspot areas. The radio spectrum is supposed to be shared between multiple AeNBs or TeNBs. The architecture, shown in Figure 8.3, intends to provide fast deployment and LTE accessibility during disasters or temporary events. Zhao and Grace [ZHA 14] propose a flexible topology management for the architecture proposed by the ABSOLUTE project. The topology management proposal is based on the activity in a given area, the greater the traffic, the better the coverage of the area. They also propose to use the satellites information to provide the first hint of where to deploy the available AeNBs and TeNBs. After that, the algorithm positions the stations based on the probability of messages retransmissions. If a given TeNB becomes inactive, below a given threshold, it is removed from that low use area and deployed into another.

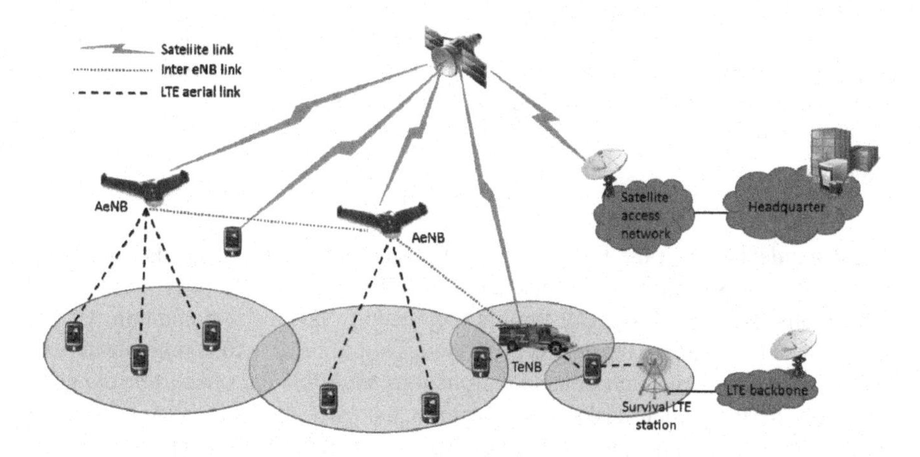

Figure 8.3. *ABSOLUTE project's main defined scenario components and their relation*

One of the main challenges of autonomous drones is to ensure connectivity even in the absence of central infrastructure. Indeed, an important task of the network consists of organizing geographically dispersed nodes to determine an appropriate topology over which high-level routing protocols could be implemented. In Section 8.8 and 8.9, we will concentrate on a specific heuristic-based approach to exemplify how this can be done. However, many others heuristics could be used to organize the network. Reynolds [REY 87] present a model to control flocking particles. This is the first and one of the major studies on distributed agent-based behaviors, a pioneering study on how bio-inspired cooperative movements can be used to control agents. Giving its decentralized approach this technique could be used as a basis to control a moving drone topology. An adaptation of Reynolds approach as already been used to control real flying drones [HAU 11].

Potential field techniques are widely used in robotic applications [KHA 86] for tasks such as goal seeking and obstacle avoidance. Attractive fields can represent a goal-seeking behavior, and repulsive fields a collision, or obstacle, avoidance behavior. Another heuristic method that could be used to provide a distributed safe-distance coordination among a group of autonomous flying drones is the virtual spring model proposed by Daneshvar and Shih [DAN 07]. According to this model, nodes in a given neighborhood radius are connected by a virtual spring. As the vehicle changes its position, speed and altitude, the total resulting forces on each virtual spring try to equal zero by moving to the mechanical equilibrium point. The agents then add the simple total virtual spring constraints to their movements to determine their next positions individually. To put this method in place, flying vehicles need no direct communication with each other, require only minimum local processing resources, and the control is completely distributed. However, this is under the assumption that each vehicle knows the neighbors' position, which in reality should come from the messages exchanged with neighbors.

8.5. Drone architecture

Independently from the kind of applications/missions that will make use of the drone airborne structure, we consider that, for PSNs, the nodes need to be autonomous, auto organized and for that they need to be able to communicate with one another. Each drone, regardless of its type, should be able to communicate with others and autonomously coordinate their actions to divide the tasks to be done. To efficiently perform their tasks, we consider that each node is organized in an architecture similar to the one described in Figure 8.4. Even though the implementation may change, to consider the specificities of each drone, it is expected that each function described in Figure 8.4 needs to be implemented for each drone [TAN 16].

In the basic architecture the role or each component is follows, the MAC (Medium Access Control) layer provides the network abstraction to all the other modules. It hides the specifics of the network technology to be used and can be interchanged to evolve or adapt to local regulations and standards. The radio management sub-system is responsible for controlling the power of the radio and optimizing communication with the other drones. The self-organizing network module is responsible for exchanging messages with the nearby drones and for coordinating efforts. The information relaying module is responsible for receiving data from the other drones and either forwards it to the next drone in the direction of the destination, or relays the data until the drone finds either the destination, or some other drone that is going in its direction.

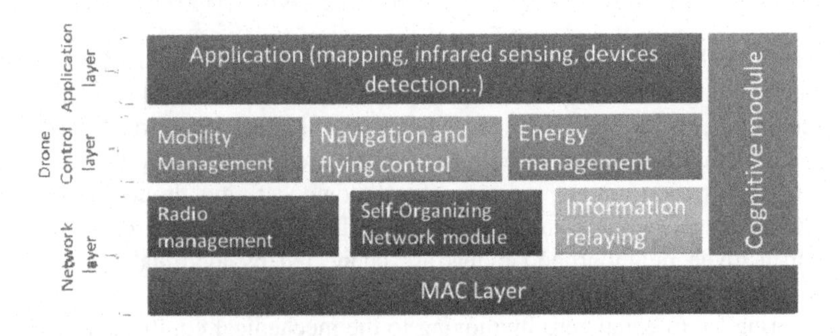

Figure 8.4. *Layered organization of modules common to all drones, independently of their type*

The mobility management module is responsible for planning the mobility of the drone, considering the objectives and the probable actions of the other drones on the region. The navigation and flying control module is responsible for implementing the planning done by the mobility management module. Based on geographic information, e.g. Global Positioning System (GPS) data, it controls the route and the power of the engines. This is the component of the architecture responsible for controlling the flight direction and speed. The energy management module is in charge of keeping track of the remaining energy and warning when it is time for the drone to return to the base. If the energy gets critically low, this module starts the emergency procedure. The emergency procedure, among other actions, consists of sending a distress message, with the current position, safely landing the drone and repeating the distress message at regular intervals.

The application layer is dependent on the task to be done at the time, the kind of drone and the type of sensors available. The application should also be interchangeable, since the tasks for the drones may evolve during the rescue operation effort.

The cognitive module, vertical to all the others (see Figure 8.4), provides generic AI algorithms that can help with the decision-making activities of all the other modules. For example, the mobility management module may use it to infer the actions, and other drones will use it to optimize the coverage of the area. The energy management module may use it to decide the best moment to return to the base. That is, based on the energy consumption, how much energy the drone should spend to fly from the point it is at to the base.

8.6. Fleet architecture

Each UAV, taking into account the work it may perform and how this work is performed nowadays, represents by itself a great help to relief effort activities. However, to be effective, and fully explore the drones' potential, more than one kind of drone would be required on disaster sites. This section proposes an autonomous communication, and coordination, architecture to enable the efficient use of different kinds of drones, and their specific characteristics, over disaster scenarios. An example of architecture is shown in Figure 8.5. In this example, three different drones perform specialized tasks adapted to their physical characteristics. The main target for this organization is to help rescue teams over the first hours after the crisis outbreak. The three categories of drones we target are: blimps, fixed wing (horizontal axis) and rotary blade (vertical axis) drones. We consider that drones are constantly aware of their autonomy and the energy required to reach the closest recharging station. When a drone observes their amount of remaining energy to be close to the limit, if warms the nearby drones that it needs to leave the formation. This warning also implies that, depending on the activity it is performing, someone else will need to take over its duties. Nearby drones will then organize themselves, and, if required, elect the one who will take over the interrupted task. This is a basic principle and applicable for all three types of drones and their specific tasks.

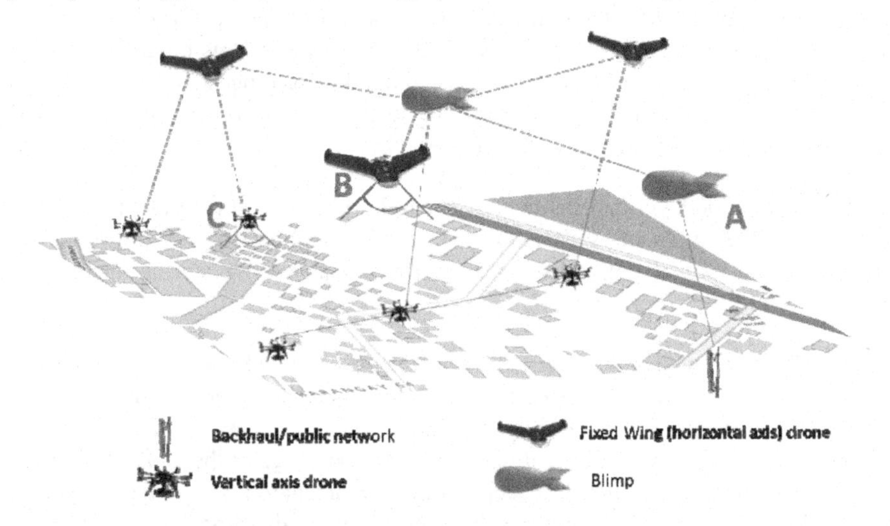

Figure 8.5. *Example of a multi-layer organization of a drone fleet, highlighting the instantaneous communication links. (**A**) Blimps that can have higher autonomy and stability. For these reasons, blimps are used to form a communication backbone. Fixed wing drones also present a good energy/speed trade-off and are being used to provide opportunistic communication links. These drones can fly at relatively high altitudes, which means their vision angle (**B**) is larger than that of vertical axis drones (**C**). Finally, we have the vertical axis drones that, despite having a lower energy efficiency, have better maneuverability and can adapt their speed and altitude to the characteristics of the terrain*

One of the main tasks drones may perform is to provide a temporary communication infrastructure. Stable and less energy-hungry devices should compose the main backbone infrastructure. Ideally, blimps could perform this role quite well. They are stable and, given their characteristics, present a much larger autonomy than the average drone. For example, the X-Tower from Stratxx can fly for 23 days broadcasting 12 channels of digital TV during all this time [STR 14]. The advantage of blimps, apart from their energy efficiency, is that they may fly at a relatively high altitude providing a large coverage footprint. Moreover, if the footprint is not enough, the covered area can always be extended by interconnecting multiple blimps. To avoid interferences, backbone nodes should have two interfaces, one to work as an access point to other nodes, and a second interface dedicated to handle the backbone traffic, i.e. routing other backbone nodes traffic and accessing the backhaul.

Fixed wing drones have a lower autonomy than blimps, but they can cover a given region much faster. They are useful for missions where one needs to cover an area and collect information about it. They can coordinate themselves directly through the blimp backbone or even through the ground backhaul, if they are in its communication range. This coordination is important for ensuring full and optimal coverage of the region by the fixed wings drones. Given its flying characteristics, the most rational way to divide the area for fixed wing drones is in strips (see Figure 8.6).

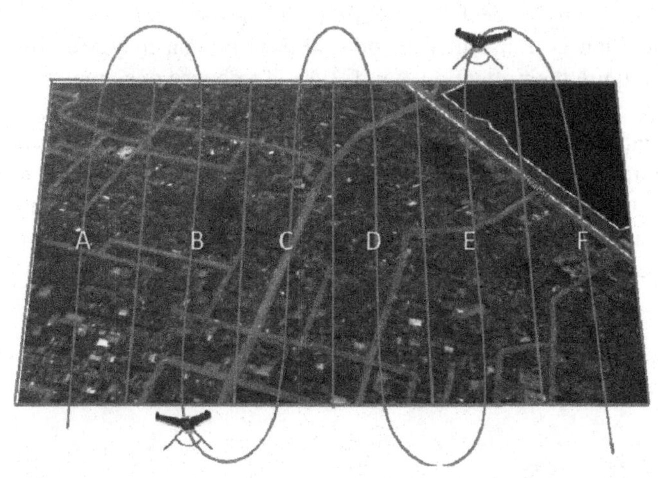

Figure 8.6. *Example of a region mapping performed by fixed wing drones. For this kind of UAVs, it makes sense to divide the target area into strips so that they can go over the sub-areas one by one doing only one turn at the end. Another advantage of dividing the mission this way is to be able to parallelize the work by breaking it into sub-tasks and distributing these over different drones, if these are available*

Vertical axis drones typically present smaller autonomy than blimps and fixed wing drones. However, their advantage is to be able to fly in every direction, horizontally and vertically, as well as hover in a fixed position. They can also steadily fly close to the ground, which could, for example, be used to provide high fidelity data from Ground Penetrating Radars (GPR) and the sensing of weak portable device signals. These can provide hits to the teams on the ground, indicating areas where they may wish to focus research efforts. Given its characteristics and limited autonomy, to be effective, horizontal blade drones should concentrate their efforts on specific interest spots, e.g. collapsed buildings, avalanche and mudslides runout zones. The interest spots could be provided by the rescue teams or these could even be automatically detected by analyzing the high fidelity multimodal maps (visible spectrum, infrared, lidar data, among others) created by the fixed wing drones. It is also important for these drones to coordinate themselves in order to provide a full, non-overlapping, coverage of the target area. As shown in Figure 8.7, taking into account the high fidelity maps, we can automatically generate a grid attributing an identification to each zone, e.g. D2, after covering a zone the drone can warn others about it, transmit the data to the operation center and reserve another non-covered area nearby. The advantage of this technique is that areas with low or no interest, e.g. G1 in Figure 8.7, can be scanned much faster than other areas, e.g. C2 which has much higher density of interest spots and can even be subdivided. The main idea is to fairly divide the work to be done, not the area itself. If drones decide to subdivide their zones into a smaller granularity grid, this decision is transmitted to other drones so that they are now aware that other areas require their attention, e.g. C2.A3. To choose one area, drones can base their decision on a number of factors that may include: distance from the present position flight plan amount of interest spots available energy level and other drone probable routes. If a drone is forced to leave an area before being able to fully cover it, e.g. due to low energy, it divides the area and warns other drones about the sub-areas it has already covered. The remaining areas go to the "work to be done" pool and will be covered by other drones later on.

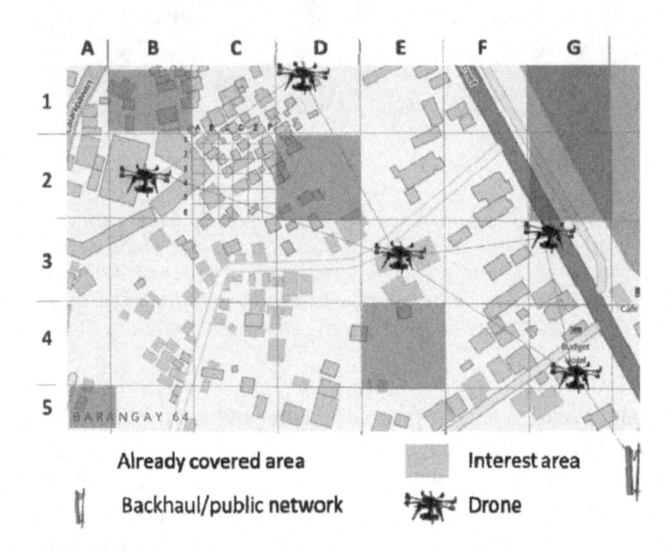

Figure 8.7. *The sensing drones can divide the target area into smaller sub-objectives. Each one of these is covered by one drone. The negotiation over areas that were already covered and those which are to be covered should be done autonomously and locally between the drones*

Data transmission may be expensive in terms of energy, therefore even if important information has to be sent immediately to the operation center via backhaul, public network or the drone's ad-hoc network, part of the data can be filtered to save energy. General collecting data could be stored in the drones and then wait to offload until the nodes return for charging. This way drones would work as mules [SHA 03] working in a collect, store and deliver way. Other data, such as self-organization controlling messages can be even confined to the drone network only.

Drones should be as autonomous as possible, but also should be able to receive instructions from distance. For example, the operation center may require an online video transmission of a specific interest area. In this specific case, as the information may be urgent, the data can be sent directly, or if the drone is not in the range of the backhaul, the video stream could be sent to the operation center in a multi-hop way. Assuming that during the transmission the drone loses its connection, the data is relayed and send to the operation center opportunistically the next time the drone is connected, working in a collect store and forward way. We cannot trust that the network will always be fully connected, so store-and-forward mechanisms [POL 11] should form part of the basic communication mechanisms.

8.7. Topology control requirements for network reliability

The process of building and maintaining a given network topology is called the topology management or topology control (TC). It comprehends the aspects related to dynamicity of the nodes and the control of transmission power to save energy and control network density [SAN 05]. Having a well-organized and predictable network structure greatly simplifies the work of the algorithms of upper layers and the stability of the network. The algorithms may act locally, distributed, or globally, centralized. Either way topology management has a systemic impact as they organize the network as a whole. The development of self-organizing MANET-like networks requires efficient and customizable autonomous topology control mechanisms. Before proceeding to details of the proposed method, we want to present the characteristics of a topology management algorithm. To be useful, the TC algorithm should:

– control the number of nodes offering a given service/performing a specific role. For example, in an LTE network, eNBs are responsible for providing access and organizing the traffic in their cells, whereas UEs could be used as relays;

– perform the topology reconstruction to reach a stable configuration, while respecting the desired topology;

– ensure stable, or at least as stable as possible, topologies;

– produce well-balanced network topologies, where no node is overloaded.

8.8. Mission-based topology description

The topology management method presented here, even though it could be applied to any kind of robotic network, targets more specifically the control of mission-based aerial drone networks. The network nodes have a predefined mission and a squadron leader that is the responsible for the squadron. During the mission, squadron nodes work as a cluster and the squadron leader works also as the cluster head, organizing the communication inside the cluster. From here on, we use cluster head and squadron leader indistinctly, as from the viewpoint of the network their function is the same. The only difference is that the squadron leader is assigned before the start of the mission, while the cluster head has a temporary role in that the nodes may provide a connection to a subset of nodes if they move far away from the squadron leader. Each squadron has a specific mission that differs among squadrons and during the time of the operation. As example of missions we could cite, hover over specific points to provide access to the ground rescue teams or scan a delimited geographic area searching for survivals. In general, nodes in the same cluster present similar patterns and tend to be close to each other.

Each squadron has his own identifier, nodes recognize their squadron leader based on this identifier. If, during the mission, the squadron becomes split, and part of the squadron is out of the range of the leader, a new leader for the squadron needs to be elected. The new leader should be the one with more resources available at the time. When two sub-squadrons are in the communication range of each other, they should merge to save communication resources.

On the merge the original squadron leader is always preferred, or in the absence of it, in the case that we are merging two sub-squadrons that were created during the mission, the node that is serving as leader to more nodes should become the leader. In case of a tie, an election should take place and, once more, the winner should be the node with more resources. Whenever a node becomes isolated, or far from its leader, it becomes a leader and connect to the other leaders in the region. If it enters in a region covered by another leader, with a higher rank, the node gives up of being a leader and asks the high-ranked leader to join its group. Figure 8.8 shows an image extracted from the Sinalgo simulator with three defined squadrons flying together.

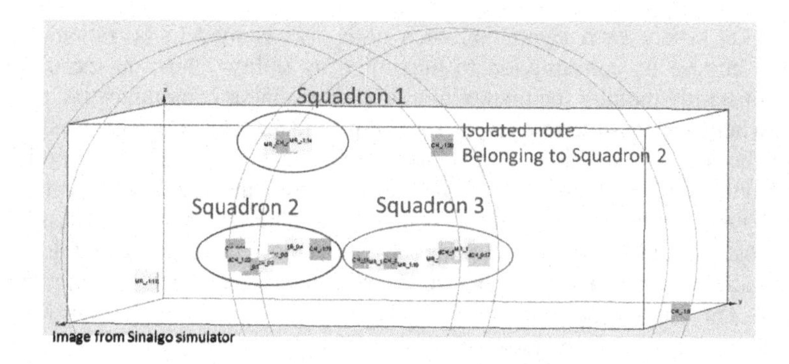

Figure 8.8. *A schematic of the simulation of an experiment with three squadrons*

8.9. Bases of the proposed method

This method uses a market-based approach where the main objective is to fulfill the requirements listed in section 8.7 to control the network described in the previous section (section 8.8). To dynamically organize the network structure, this heuristic approach uses the economic concepts of supply and demand. In his book The Wealth of Nations, Adam Smith says "It is not from the benevolence of the butcher, the brewer, or the baker that we expect our dinner, but from their regard to their own interest" [SMI 76]. As we will show, just as in Adam Smith's statement, even though nodes behave selfishly, using the marked-based approach, they manage to reach the best possible allocation, as it fits their own interests.

The three laws of supply and demand can be described as follows [LEV 06]:

1) When demand is greater than supply, prices rise and when supply is greater than demand, prices fall;

2) The greater the difference between supply and demand, the greater the force on prices.

3) Prices tend to an equilibrium point, called the Walrasian equilibrium, at which supply equals demand.

Going back to the TC algorithm requirements from section 8.7, we can map the need to control the number of clusters to the first law of supply and demand:

As in the real world Internet service provider (ISP) market, a cost is associated with the services the node requests from/provides to others, i.e. a basic communication price. At any given time, different providers may have different prices, depending on both their type and load. By controlling the prices of different services offered in the network, we can control the number of nodes offering each one. Regardless of the exact pricing scheme, the proposed mechanism emulates a free market where each agent, i.e. each node, is assumed to be rational and to selfishly choose its consumption to maximize its utility. Thus, on the one hand, nodes constantly monitor the market in search of a smaller communication price and switch whenever possible to cheaper service providers. This movement reduces the cost of the previous provider, and at the same time increases the value/cost of the second one, as the communication prices vary with the number of nodes being served by the cluster head. However, if these prices become too high, cluster heads start to lose customers out to their competitors. In some conditions, client nodes may even decide it is cheaper to become providers themselves (thereby initiating new clusters). This fluctuating prices mechanism tends to reach an equilibrium where the prices are more or less homogeneous.

This is a good news because the network is also required to converge to a stable state as fast as possible. The need for an equilibrium can be related to the third law of supply and demand, whereas the second law implies that the greater the differences between supply and demand, the greater the force on prices and the faster the resulting convergence.

In fact, any free, or competitive, market such as the one we just described, under certain conditions, leads to the Walrasian equilibrium [LEV 06]. This equilibrium point is also a Pareto-optimal arrangement, where no changes on the allocation of goods and services can beneficiate a participant without causing damage to others. This follows from the first theorem of welfare economics [DEB 59]:

– [A1] The market for all possible goods exists and there are no externalities present, i.e. all costs and benefits are transmitted through prices. In this model, the only price is the one given by a public pricing formula.

– [A2] The market is perfectly competitive and no participant has enough power to influence the prices. This is also satisfied as the costs vary and if they get too high even the clients themselves may become providers, with this mechanism the method breaks any potential monopoly.

– [A3] The cost of transactions is negligible. There is no hidden cost attached to the transactions.

– [A4] Market participants have perfect information, all agents are rational and have access to full information about all products at all times. This is ensured by having prices for every possible provider, and node type, frequently exchanged among the nodes.

Thus, a globally fair and efficient allocation of resources and the corresponding equilibrium point is achieved, in a competitive market, when supply equals demand for any good or service traded among the peers, in our case, the connection/communication with the rest of the network. The Algorithm 1 presents a broad view of the protocol put in place for controlling the topology of the nodes in the experimentation section. All nodes, apart from the squadron leaders start as Isolated

Nodes (IN). These IN try to find a leader (Cluster Head – CH), from the same squadron, if they find one they attach to this leader, if not they become one. Nodes have two interfaces, one to talk to the nearby nodes in the same group. The second interface is used only by the group leaders to connect to other group leaders. This second interface has a larger range and intends to interconnect the various clusters, thus it is accessible to all nodes independent of their squadron and size.

8.10. Experiments

The evaluations were carried out using the Sinalgo simulator [SIN 07] in a 3000 × 3000 × 2000 m space where nodes ran for two simulated hours. The simulations were conducted with 400 and 700 m range for interface one and two, respectively. All experiments were conducted using Linux Ubuntu 14.04.1 LTS on an Intel Xeon W3670 3.20 GHz twelve cores machine with 12 GB of RAM. All graphs are presented with a confidence interval of 99% and each point is the result of the averaging over at least 34 runs with different network configurations. The nodes arrive randomly and are placed near to their squadron leader, nodes have the tendency to follow their leaders, but this tendency is not fully enforced by the mobility model. For the experimentations, nodes fly over the area following a normal random way point speed distribution where the average is 10 km/h and on average nodes maintain one direction for at least 20 s.

Nodes are divided into squadrons, regular nodes (Mobile Routers – MR) from different squadrons do not talk directly to each other. The communications need to go through the cluster heads, which can transfer messages through the second interface. Table 8.1 summarizes the main simulation parameters used for the experiments.

Parameter	Value
Simulator	Sinalgo
Simulation time	2 h
Area (L × W × H)	3 km × 3 km × 2 km
Interface 1 range	400 m
Interface 2 range	700 m
Percentage of loss messages	0.01%
Average speed	10 km/h
Number of squadrons	1, 2, 3, 4, 5
Prices to connect	DH:0, CH:5, MR:15
Max cluster size	16 nodes

Table 8.1. *Main simulation parameters*

```
1.  The node arrives in the network (IN, mySquadron); //isolated node, present squadron identifier
2.  The node broadcasts a Connection request (mySquadron) message;
3.  Waits for responses;
4.  If (receives any Connection response from a possible provider) {
5.      Weights the costs of the responses;
6.      Sends a Connection confirmation to the provider with the lowest cost;
7.      Becomes a MR; // Mobile router
8.      Go to step 17
9.  } Else{
10.     If(the number of trials smaller than 3){
11.         Returns to Step 2;
12.     }Else{
13.         Becomes an CH; //Cluster Head
14.         Sends a Connection Update;      }
15. }
16.  Starts the broadcast timer;
17.  Starts the evaluate_update timer;
18.  Triggers TreatBroadcastTimer();
19. While(running){
20.     Waits for messages or broadcast timer to expire;
21.     If(broadcast timer expired)
22.         Calls TreatBroadcasTimer();
23.     If(a Connection Request is received and squadronID == mySquadronID){
24.         Answers with a Connection Response, informing present connections;
25.     }Elseif(a Connection Confirmation is received){
26.          Registers the connection;
27.          Reevaluates present state {
28.               If (is a MR) {
29.                    Becomes a CH;
30.                    Sets the price to the basic CH price;
31.               } Elseif (is a aCH)
32.                    Increments the price;
33.          }
34.     }Elseif(a Connection Response is received){
35.         If( the response node cost is lower than the present one){
36.              Sends a Connection Confirmation;
37.              Registers the received Update
38.              Registers the new connection;
39.              Reevaluates state (if a CH, becomes a MR);
40.              Sends a Connection Cancel present CH;
41.         }
42.     }Elseif (a Connection Update(squadronID) is received){
43.         If(squadronID == mySquadronID)
44.              Registers the received Update, ;
45.     }Elseif(a Connection Cancel is received){
46.         Removes the related connection;
47.         If(the removed connection is to the current provider (CH)){
48.             Becomes an IN;
49.             Returns to Step 1;            }
50.     }
51.     If(evaluate_update timer expired){
52.         Evaluate Updates to find better providers;
53.         If(found better provider)
54.              Sends a Connection Request(mySquadronID);          }
55. } // While
56.  function TreatBroadcastTimer(){
57.     If(state is CH){
58.            Broadcasts a Connection Update(mySquadronID).
59.            Returns to Step 17;       }
60. }
```

Algorithm 8.1. *High-level algorithmic description of the market-based topology control for aerial drones. The mysquadronid variable is a predefined squadron identifier*

The main objective of the experiments is to show that the market-based approach can effectively be used to control the topology aerial drones. Here we will use the market topology control to organize the network in a two-layer hierarchical structure, but other kinds of organizations are also possible. The graph in Figure 8.9 presents the number of cluster heads when we vary the density of nodes and the number of independent squadrons in the same area. We can perceive that the number of cluster heads, the nodes responsible for organizing the network is linked to the number of nodes in the network. We can also perceive that the greater the number of squadrons the greater the number of cluster heads. This is expected as the nodes can only exchange information and connect to clusters within the same squadron. The larger the number of squadrons, the smaller the probability of finding a cluster around on the same squadron, thus the number of cluster heads, to manage the growth of different squadrons.

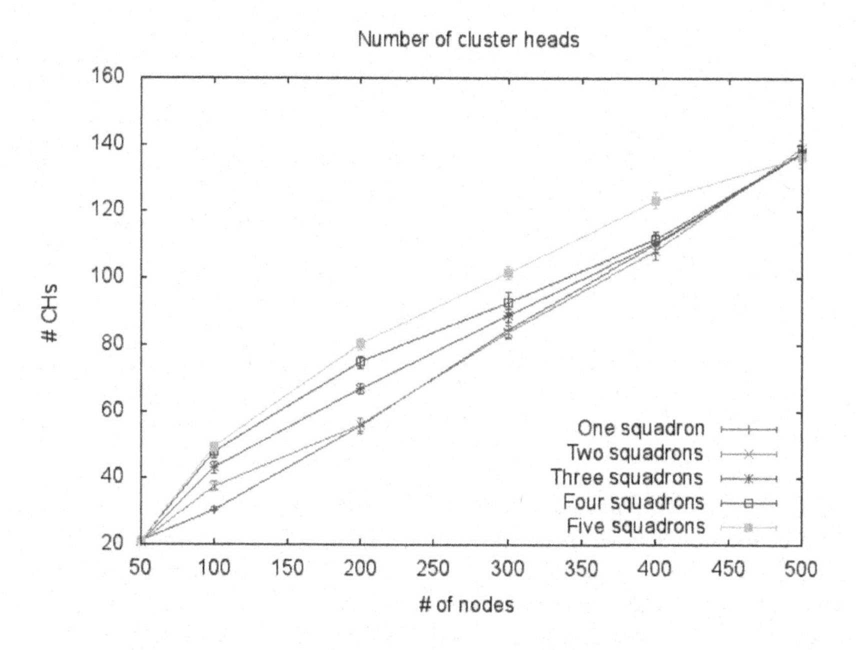

Figure 8.9. *Average number of cluster heads over the network. For a color version of the figure, see www.iste.co.uk/camara/wireless3.zip*

Figure 8.10 presents the average size of the clusters, we can perceive that they stabilize around 3.5 nodes per cluster. This is linked to the used mobility model, nodes when flying together tend to form stable clusters, but as the random direction mobility model does not enforce the formation, clusters tend to split and nodes change from one cluster to another. We can also observe clear influence of the number of squadrons over the size of the clusters. The bigger the number of squadrons the smaller the cluster sizes.

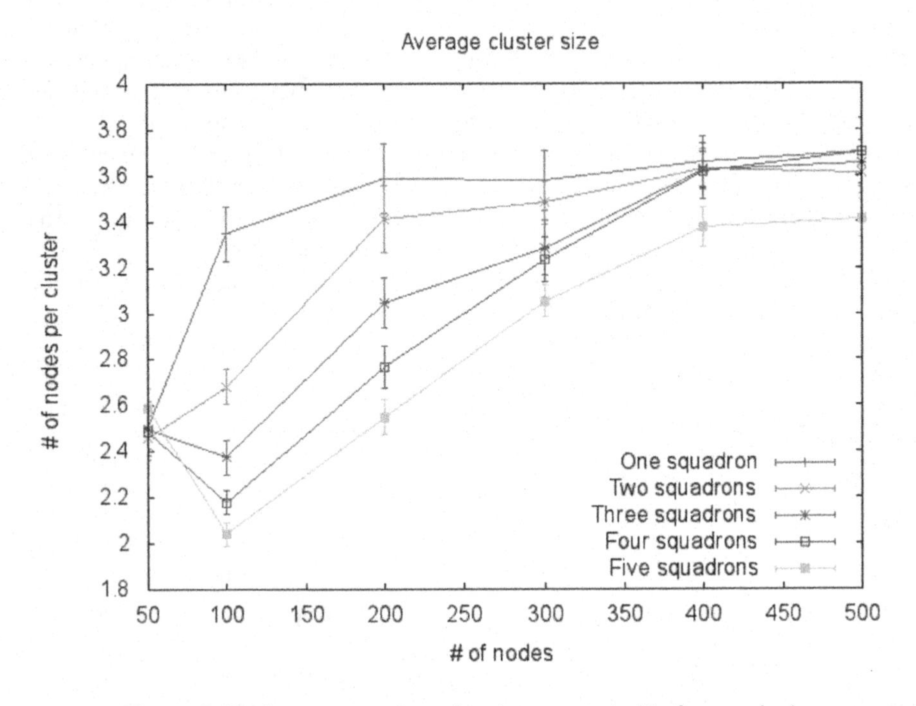

Figure 8.10. *Average number of nodes connected to form a cluster during the simulation period. For a color version of the figure, see www.iste.co.uk/camara/wireless3.zip*

With the mobility, part of the nodes tends to become CH; however, they do not provide connection service to no other node, i.e. no other node attaches to them. This is a regular and expected behavior; however, it should remain low, since these stand-alone cluster heads represent a cost in terms of control messages, which helps to decrease the network autonomy and hardens the access to the medium. Figure 8.11 shows the average number of stand-alone CHs over the whole simulation time, taking into account the size of the network and the number of squadrons. For small densities, the number of stand-alone CHs tends to be stable, and linked to the number of squadrons acting in parallel over the target area. With the increase in the network density, nodes also tend to spread and move more over the target region and this increases the number of nodes that become CH to try to maintain the connection with the rest of the network.

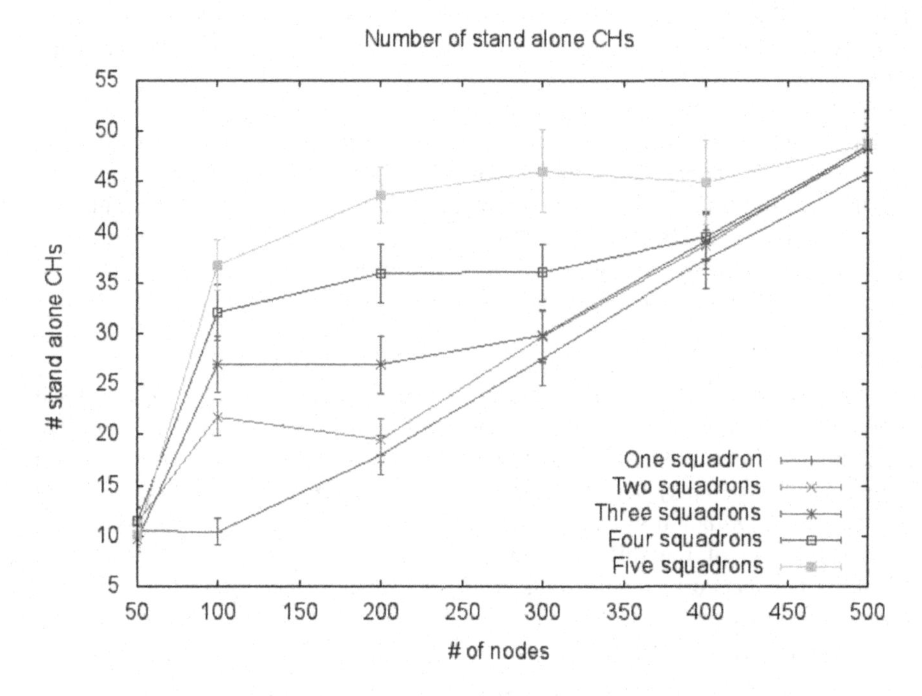

Figure 8.11. *Average number of stand-alone clusters during the simulation period. For a color version of the figure, see www.iste.co.uk/camara/wireless3.zip*

8.11. Conclusion

Surely enough, search and rescue operations can make use of drones for a fair amount of activities. However, if not autonomous and capable of self-organizing, these elements can be more of a burden than a help in a catastrophe scenario. Rescuers must focus on the activity they have at hand, i.e. saving lives. It is not productive to make them spend their time handling drones and their tasks. The proposed architecture intends to provide the organization required for a fleet of drones to autonomously, in a push-button way, scan the region and provide useful information. Another intention of the proposed architecture is to use this fleet to provide communication over disaster areas, even for severely affected areas.

It is also important to note that drones should be able to perform opportunistic communication and coordinate with the nearby nodes. In a disaster scenario, store-carry-and-forward techniques may be the only way to convey important information among the computational elements. Drones can exchange information with each other about the route, and strategies, they are taking and if they are moving in the direction of the operation center, they can carry the messages of other drones until their final delivery at the destination. In the same way, message ferries are designed [POL 11].

Different kinds of drones may provide different services and, ideally, should play the roles they fit the best. Even though we could exchange some of the tasks among the different drones, it would have an impact on the end results. For example, we could, without a doubt, use fixed wing drones to create a mobile backbone. However, not only would the organization of the drones to provide constant full coverage be more complex, but also the lifetime of the backbone would be much smaller.

In a disaster scenario, to be effective, rescuers require a series of information and services, for example up-to-date geographical information and network connection capabilities, among others. Aerial drones may fulfill part of these needs. In the future, aerial drones may become an invaluable tool to provide a number of different services and information to help on disaster relief efforts. However, to be effective, the organization of the different kind of nodes needs to be as autonomous as possible. The network should work with the least human interaction possible. Among the tasks that should be accomplished autonomously by the nodes is the auto-organization of the network.

This chapter discusses the problem of creating and maintaining a stable network topology for aerial drone networks. We present different initiatives and methods, and we discuss how a market-based heuristic can be used to control the topology of a large aerial drone network. The presented method, which is based on the way the economics principles of offer and demand, successfully organizes the topology even in the presence of honest selfish nodes. We call them honest because they do not try to cheat their prices to attract/refuse connections. On the other hand, they are selfish because they try to get the best connection cost possible, and for this reason they are constantly searching for "best deals" among the clusters in the region of interest.

The network tends to a stable point where all the clusters have more or less the same price, and where it is not worth changing providers. However, as the prices are linked to the number of nodes being served by a cluster, and nodes move, leaving old clusters and connecting to new ones, the prices vary and nodes enter in concurrency. It is interesting to notice that even in a selfish environment we can foster collaboration. In this competitive environment, we only set the basic prices to different types of services and we let the nodes decide whether they are willing to pay the price or not. Nodes do not collaborate to form a stable and consistent topology because they are altruistic. They collaborate because they can gain something with this collaboration, in this case, "pay" less for their connection. This simple fact, instead of penalizing collaboration fosters it. Nodes now have a reason to collaborate and even to behave well to each other. The topology is relatively stable, but more experiments need to be made in order to evaluate the impact of the technique with more realistic mobility models and different mission-based scenarios.

8.12. Bibliography

[AME 15] AMERICAN RED CROSS AND MEASURE, Drones for Disaster Response and Relief Operations, Report, April 2015.

[ARS 09] ARSLAN O., INALHAN G., "Design of a decision support architecture for human operators in UAV Fleet C2 applications", *14th International Command and Control Research and Technology Symposium (ICCRTS)*, Washington, DC, June 2009.

[CUM 07] CUMMINGS M.L., BRUNI S., MERCIER S. *et al.*, "Automation architecture for single operator, multiple UAV command and control", *The International C2 Journal*, vol. 1, no. 2, pp. 1–24, 2007.

[DAN 07] DANESHVAR R., SHIH L., "Virtual spring-based 3D multi-agent group coordination", *International Conference on Complex Systems (ICCS 2007)*, Boston, MA, 28 October–2 November, 2007.

[DEB 59] DEBREU G., *Theory of Value: An Axiomatic Analysis of Economic Equilibrium*, John Wiley & Sons, 1959.

[DEV 13] DEVAUL R., TELLER E., BIFFLE C. *et al.*, Balloon network with free-space optical communication between super-node balloons and RF communication between super-node and sub-node balloons, Patent PCT/US2013/020705, 18 July 2013.

[FAC 14] FACEBOOK, "Connecting the World from the Sky", available at: https://info.internet.org/fr/blog/2014/03/28/connecting-the-world-from-the-sky/, last visited 14 December 2016, 2014.

[FP7 12] FP7 ABSOLUTE, "Aerial Base Stations with Opportunistic Links for Unexpected & Temporary Events", available at: http://www.absolute-project.eu, last visited 7 July 2015, 2012.

[GAR 14] GARSIDE J., Facebook buys UK maker of solar-powered drones to expand internet, The Guardian, available at: http://www.theguardian.com/technology/2014/mar/28/facebook-buys-uk-maker-solar-powered-drones-internet, last visited 7 July 2015, 28 March 2014.

[GIL 14] GILMAN D., EASTON M., Unmanned Aerial Vehicles in Humanitarian Response, Policy Paper, OCHA Policy and Studies Series, June 2014.

[GOO 15] GOOGLE, "Balloon-powered Internet for everyone", available at: http://www.google.com/loon/, last visited 7 July 2015.

[HAU 11] HAUERT S., LEVEN S., VARGA M. *et al.*, "Reynolds flocking in reality with fixed-wing robots: communication range vs. maximum turning rate", *IEEE/RSJ International Conference on Intelligent Robots and Systems*, San Francisco, CA, 25–30 September 2011.

[HUM 16] HUMANITARIAN UAV NETWORK, available at: http://uaviators.org/, last visited 15 October 2016, 2016.

[ICA 16] ICARUS, ICARUS Public Report summarizing the project outcomes, available at: http://www.fp7-icarus.eu/icarus-public-report, June last visited 15 October 2016, 2016.

[INT 16] INTERNATIONAL CHARTER, "Space and Major Disasters", available at: https://www.disasterscharter.org/web/guest/home, last visited 15 October 2016, 2000.

[KHA 86] KHATIB O., "Real-time obstacle avoidance for manipulators and mobile robots", *International Journal of Robotics Research*, vol. 5, no. 1, pp. 90–98, 1986.

[KRI 09] KRISHNAMURTHI N., GANGULI A., TIWARI A. *et al.*, "Topology control within the airborne network backbone", *Military Communications Conference, MILCOM* 2009, Boston, MA, 18–21 October 2009.

[KWO 15] KWONG M., "Nepal earthquake: drones used by Canadian relief team", CBC News, 27 April 2015.

[LEV 06] LEVIN J., General Equilibrium, Harvard University 2006.

[MAD 15] MADRY S., *Space Systems for Disaster Warning, Response, and Recovery*, Springer, 2015.

[MEI 14] MEIER P., "Humanitarians in the Sky: Using UAVs for Disaster Response", available at: http://irevolution.net/2014/06/25/, last viewed 30 June 2014, 25 June 2014.

[MID 02] MIDKIFF S.F., BOSTIAN C.W., "Rapidly-deployable broadband wireless networks for disaster and emergency response", *Proceedings First IEEE Workshop on Disaster Recover Networks*, June 2002.

[MIL 03] MILNER S., THAKKAR S., CHANDRASHEKAR K. *et al.*, "Performance and scalability of mobile wireless base-station-oriented networks", *ACM SIGMOBILE MC2R*, vol. 7, 2003.

[PED 14] PEDERSEN J., "Use of UAVs in the NGO World", *CRS Conference – ICT4Development*, Nairobi, Kenya, 25–28 March 2014.

[POL 11] POLAT B.K., SACHDEVA P., AMMAR M.H. *et al.*, "Message ferries as generalized dominating sets in intermittently connected mobile networks", *Elsevier Journal on Pervasive and Mobile Computing*, vol. 7, no. 2, pp. 189–205, 2011.

[REY 87] REYNOLDS C.W., "Flocking, herds, and schools: a distributed behavioral model", *Computer Graphics*, vol. 21, no. 4, pp. 25–34, 1987.

[SAN 05] SANTI P., *Topology Control in Wireless Ad Hoc and Sensor Networks*, Wiley, 2005.

[SHA 03] SHAH R.C., ROY S., JAIN S. *et al.*, "Data MULEs: Modeling and analysis of a three-tier architecture for sparse sensor networks", *Ad Hoc Networks*, vol. 1, nos. 2–3, pp. 215–233, September 2003.

[SHI 11] SHIRAZIPOURAZAD S., GHOSH P., SEN A., "On connectivity of Airborne Networks in presence of region-based faults", *Military Communications Conference*, MILCOM 2011, pp. 1997–2002, 7–10 November 2011.

[SIN 07] SINALGO, Simulator for Network Algorithms, Distributed Computing Group at ETH Zurich, available at: http://disco.ethz.ch/projects/sinalgo/, last visited 10 July 2015, 2007.

[SMI 76] SMITH A., *An Inquiry into the Nature and Causes of the Wealth of Nations*, London, 1776.

[SOE 16] SOESILO D., SANDVIK K.B., "Drones in Humanitarian Action – A survey on perceptions and applications", available at: http://drones.fsd.ch/wp-content/uploads/2016/09/Drones-in-Humanitarian-Acion-Survey-Analysis-FINAL21.pdf, last visited 15 October 2016, 2016.

[STR 14] STRATXX, News Release, available at: http://www.stratxx.com/blog/archive/2014/03/19/, last viewed 30 June 2014, 20 February 2014.

[SWI 97] SWISS FOUNDATION FOR MINE ACTION, available at: http://fsd.ch, last visited 15 October 2016, 1997.

[TAN 16] TANZI T.J., CHANDRA M., ISNARD J. *et al.*, "Towards 'Drone-Borne' disaster management: future application scenarios", *ISPRS Annals of the Photogrammetry, Remote Sensing and Spatial Information Sciences,* Prague, Czech Republic, 12–19 July 2016.

[UNI 16] UNITED NATIONS, ICT in Disaster Risk Management Initiatives in Asia and the Pacific, United Nations Economic and Social Commission for Asia and the Pacific (ESCAP) Report, 2016.

[ZHA 14] ZHAO Q., GRACE D., "Dynamic topology management in flexible aerial-terrestrial networks for public safety", *1st International Workshop on Cognitive Cellular Systems*, Germany, 2–4 September 2014.

[ZUC 14] ZUCKERBERG M., available at: https://www.facebook.com/zuck/posts/10101322049893211, last visited 7 July 2015, 27 March 2014.

9

Safe and Secure Support for Public Safety Networks

The advent of autonomous vehicles and UAVs offers additional support for disaster relief efforts. These self-managing objects are able to gather data, deploy a wireless network, and other simple routine maintenance tasks, freeing up aid workers to focus their efforts on the rescue itself. Fleets of UAVs may deploy mobile communication networks, and with their mobility advantage, also gather visual images of areas to coordinate relief efforts. Autonomous vehicles may deploy a higher-powered wireless relay, and also transport supplies or evacuate patients from treacherous areas without risking the life of a driver. However, during these precarious situations, faults in the design of these autonomous vehicles, should they be exposed and exploited, may worsen the disaster. It is important to ensure that these deployed autonomous objects will execute both safely and securely. In this chapter, we present a design methodology SysML-Sec in the free and open-source toolkit TTool for the design and verification of both autonomous objects and mission planning.

Chapter written by Ludovic APVRILLE and Letitia W. LI.

9.1. Introduction

As explained by Tanzi *et al.* in the first volume of this book [TAN 15a], communicating and autonomous devices will surely have a role to play in the future Public Safety Networks. The "communicating" feature comes from the fact that the information should be delivered in a fast way to rescuers. The "autonomous" characteristic comes from the fact that rescuers should not have to concern themselves about these objects: they should perform their mission autonomously so as not to delay the intervention of the rescuers, but rather to assist them efficiently and reliably.

Previous work presented how such objects could play a role in PSN, either as an active agent providing a direct support to the network, e.g. as a transmission relay, or as a data provider fed into the network, e.g. a sensor gathering information and relying on the PSN to transmit its data.

In particular, UAVs have already been proposed as a solution to cover both situations. First, they provide a mobile communication network, usually by relying on a fleet of drones. Second, they capture information from, for example, cameras and lidars, and forward the captured information to the rescuers via the PSN [TAN 14]. Contributions in the area have underlined the fact that the drones must be as autonomous as possible for the reasons previously indicated [RAN 13].

Other communicating objects can serve as relays: remote sensors, mobile phones, emergency smart watches, smart medical appliances, etc.; a PSN will be formed out of a potentially huge set of Internet of Things dedicated to assistance in emergency situations: the Things for PSN (TPSN). TPSNs differ from public Internet of Things because TPSNs are assumed to be highly reliable and secure. Their high level of reliability is necessary due to the fact that rescuers must obtain the right information on time, and the fact that the object itself should not provoke extra damages. The security and privacy constraint is due to the manipulation of highly sensitive information during emergency situations, e.g. the information could be classified as confidential or reserved for justice procedures. Thus, they should not be accessible to journalists or to enemy governmental agencies.

Finally, TPSNs must be safe, secure and autonomous. This chapter focuses on the safety and security aspects, and exemplifies them with an autonomous system. More precisely, we present how these objects could be designed, and how their missions could easily be planned and verified prior to any intervention.

The chapter is organized as follows. In section 9.2, we present concrete scenarios where autonomous objects could play an interesting role; section 9.3 presents the running example; section 9.4 presents our modeling and validation method named SysML-Sec; then, section 9.5 presents how a mission validation can be modeled and performed with our approach; section 9.6 explains how our approach favorably complements similar contributions; finally, section 9.7 presents the perspective of our approach.

9.2. Context

This section presents the roles that autonomous objects could play in the scope of an emergency situation, and how they interact with the PSN deployed in the scope of an emergency situation. We take the example of two different complex and autonomous objects: an autonomous UAV and an autonomous car. Figure 9.1 shows a possible scenario using UAVs and an Autonomous Emergency Vehicle during disaster relief.

Figure 9.1. *UAVs and autonomous emergency vehicle support during disaster relief*

9.2.1. *UAVs*

UAVs could play a key role for the following missions described in [TAN 14]. The two first scenarios rely on PSN in order to send their data to the rescuers. In the last scenario, the UAV has a direct role on the PSN:

– detection and monitoring of people/victims impacted by the crisis. Typically, the drone has to identify groups of disabled persons, and to make a clear distinction between adults and children. Cameras are a good option to find victims. When victims are buried, e.g. when an earthquake has occurred, another option is to rely on special antennas to track the electromagnetic fields emitted by personal electronic devices, e.g. smartphones and smartwatches. Indoor navigation could also be used to find victims within a building on fire. Last, drones could be used to inform the victims about the situation, e.g. with loud audio messages;

– the continuous assessment of the situation status concerning the impacted area. This task includes the identification of best access roads to the disaster area and to the victims. In this context, "best" may refer to "safest" or "fastest";

– offering a mobile relay to the wireless section of the PSN. Contributions in the domain commonly propose using a fleet of drones in order to cover larger wireless communication areas.

9.2.2. *Autonomous cars*

Compared with UAVs, autonomous cars can carry more equipment, have a longer operational time, but are restricted to only slightly damaged areas with traversable roads. Figure 9.2 shows an Autonomous Emergency Vehicle with sensors and a wireless relay. Their use within an emergency situation includes:

– carrying material, e.g. bringing water, food or medicines to victims, bringing rescue equipment to rescuers and bringing energy to other equipment, e.g. reloading spots/charging stations for UAVs. They could also be used to carry victims from dangerous areas without risking the life of a driver;

– assessing a situation. An autonomous car can easily investigate road quality. It can also deploy its own sensors for data collection;

– serving as a mobile wireless relay. Emission power can be much higher that of an UAV, but with reduced mobility.

9.3. Case study

As our running example, we examine the design of an Autonomous Emergency Vehicle capable of the disaster relief tasks previously presented. The autonomous vehicle must coordinate with Central Commands, receiving commands regarding future destinations or operational tasks and updated road conditions, while sending gathered data concerning current status of the affected area.

In addition, the Autonomous Emergency Vehicle must navigate to destinations, calculating routes and vehicular operations based on incoming sensor data, such as braking and steering. We present its design in the context of the stages of the SysML-Sec methodology in the following section.

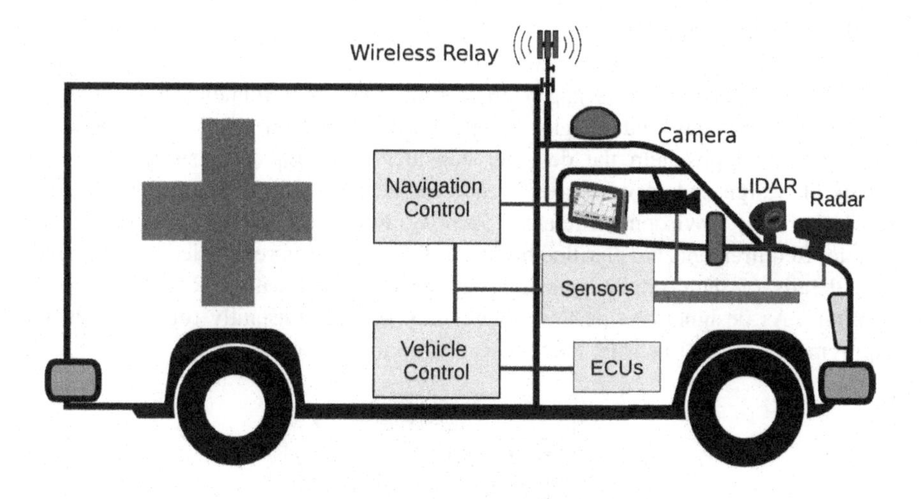

Figure 9.2. *Autonomous emergency vehicle*

9.4. Our approach: SysML-Sec

9.4.1. *Methodology*

The SysML-Sec methodology, summarized in Figure 9.3, addresses all stages in the design of an embedded system with simulation and verification at each stage [ROU 13]. We present an overview of all three stages before describing the partitioning stage in detail.

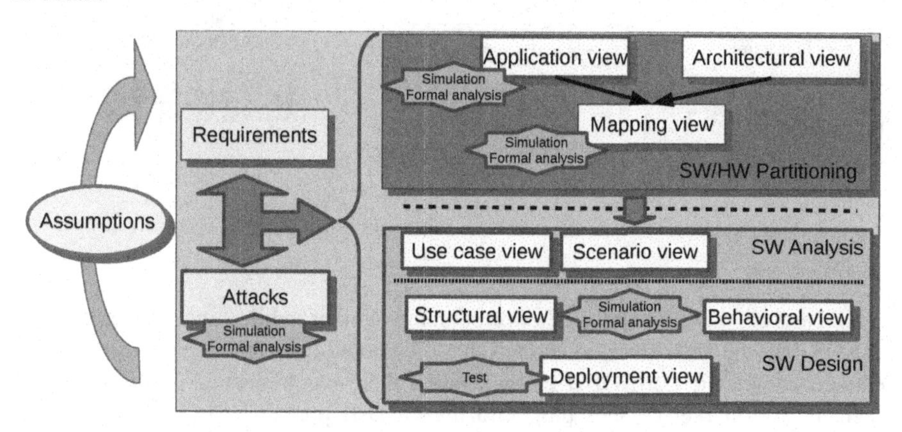

Figure 9.3. *Overall SysML-Sec methodology*

9.4.2. *Analysis stage*

The *Requirements/Attacks* stage (left section of Figure 9.3) intends to identify and analyze requirements and attacks together with the main application functions. Requirement graphs help the designer consider the complete system in the first phases of design and serve as a reference for design teams. Figure 9.4 shows an excerpt of the Environment-related Security Requirements for an autonomous vehicle. Requirements are clarified by being divided into more detailed requirements, until details of their implementation may be described with a << deriveReq >> operator. As designers refine the model, they should constantly refer back to the requirements graphs to ensure they have adhered to the standards.

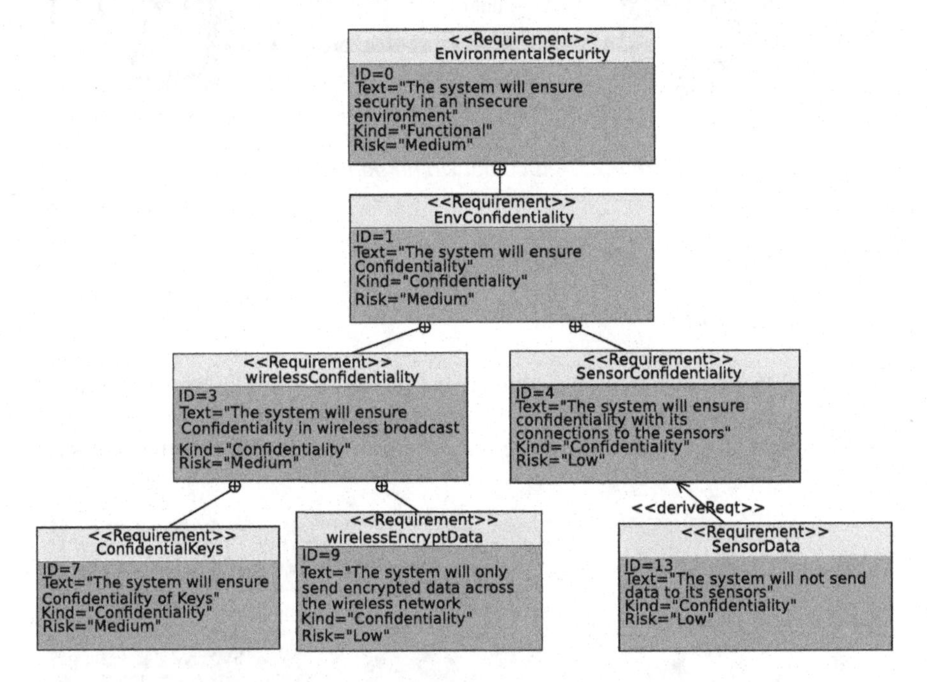

Figure 9.4. *Extract of autonomous vehicle requirements diagram*

Researchers have recently discovered many vulnerabilities in connected cars. In [KOS 10], the authors demonstrated that they could control a vehicle through the Onboard Diagnostic Port (OBD-II) by sending their own generated CAN messages. As many aftermarket devices plug into the OBD-II port and can be managed by smartphone, the compromise of these devices could allow an attacker to remotely control the vehicle. The authors later [CHE 11] presented other vulnerabilities, such as a buffer overflow bug which could be exploited so that playing a WMA file would send CAN packets. Most notably, Miller and Valsasek [MIL 15] used a vulnerable Internet connectivity feature to modify firmware and send forged CAN messages to remotely control the vehicle. This attack required no physical access to the targeted vehicle, as it could attack any vehicle on the Sprint network.

The additional connectivity required by autonomous vehicles provides even more avenues of attack, so the consideration of these attacks is vital to the design process. We capture attack scenarios (which exploit combinations of vulnerabilities) with formally defined attack graphs [APV 15]. Once defined, these graphs can be easily migrated for reuse in analysis of other systems. Attacks can be linked together in order to assess the impact of a specific vulnerability and the need to address it at the risk assessment phase, e.g. once a mapping is under evaluation.

Threats are displayed in blocks representing the target of the attacks, better presenting the attacks in the context of assets. Attacks (<< attack >> stereotype) can be linked together with primitive operators. These operators are either logical operators like AND, OR and XOR, or temporal causality operators like SEQUENCE, BEFORE or AFTER. Temporal constructs describe the attacker's operational point of view in embedded systems, in situations where there is a maximum duration between two causally related attacks. For example, when attacking a system with time-limited authentication tokens, the token must be first retrieved, and then the use of this token must occur before its expiration.

Attack instances in different parametric diagrams can be linked together in order to assess the impact of a specific vulnerability and the need to address it at the risk assessment phase. An attack can also be tagged as a root attack, meaning that this attack is at the top of a tree of attacks. Attacks can be linked to requirements, thus allowing an automated check of the coverage of attacks by verifying whether each attack is linked to at least one security requirement.

Figure 9.5 shows a sample attack of recovering GPS data regarding previous destinations, violating user privacy. We predict that the attacker could target either the GPS Gateway or Navigation Control. For example, to succeed in the attack via the GPS Gateway, the attacker must send a command to the GPS to access previous locations, send the data to the attacker and also interpret or decrypt the received GPS data. The details of individual attacks may be presented in their own attack graphs, such as the steps to gain access to a Navigation Control messaging system. As the system is developed, designers may correct vulnerabilities and then indicate certain attacks are impossible. Afterwards, formal verification determines which root attacks are still possible in the system.

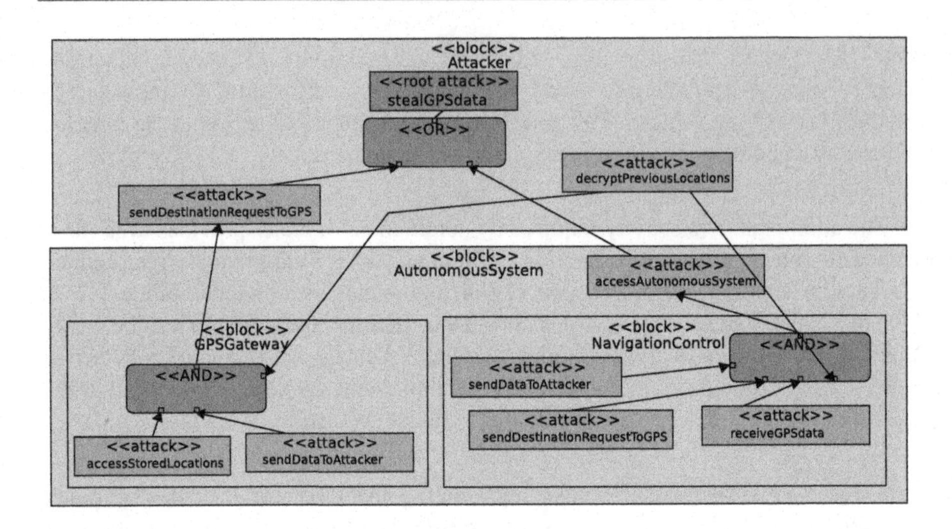

Figure 9.5. *Sample attack modeling*

9.4.3. *Hardware/software partitioning*

The *Hardware/Software* Partitioning stage (upper right of Figure 9.3) intends to determine what is the best way to implement the embedded system. For example, from a flexibility point of view, software is preferable to hardware. On the contrary, from a power consumption point of view, hardware should be used first. Thus, designing a complex embedded system, such as the embedded system of an autonomous car, requires the use of this methodological stage. Usually, a partitioning follows the Y-chart approach (as shown in the upper right section of Figure 9.3), first modeling the abstract functional tasks (application view), candidate architectures (architectural view) and finally mapping tasks to the hardware components (mapping view) [KIE 02]. Simulation and verification techniques are applied at the mapping stage to decide whether the selected architecture complies with the system's requirement. The HW/SW Partitioning phase of SysML-Sec follows this approach.

9.4.3.1. *Application view*

The Application View comprises a set of communicating tasks. The behavior of tasks is described abstractly. Functional abstraction allows us to ignore the exact calculations and data processing of algorithms, and consider only their relative execution time. Data abstraction allows us to consider only the size of data sent or received, and ignore details such as type, values or names. On the Component Design Diagram, an extension of the SysML Block Instance Diagram, the designer specifies the list of tasks, and within the task, attributes and ports indicating communication. Purple ports indicate an event-based communication and blue ports indicate a channel-based communication. Events notify another task about an event, such as the start of a function or a hardware interrupt. Channels are used for the transfer of data. Activity diagrams are used to give a behavior to tasks.

Figure 9.6 shows the Component Design Diagram of the Autonomous Vehicle in our case study. The design includes a main task "Navigation Control" communicating with "Sensor Gateway" managing all sensor data, "GPS Gateway" managing location data, "Command Gateway" managing commands of new destinations or tasks from Central Command. Based on the input data, Navigation Control calculates a trajectory and sends the driving commands to "Vehicle Control Gateway" which then interfaces with the ECUs.

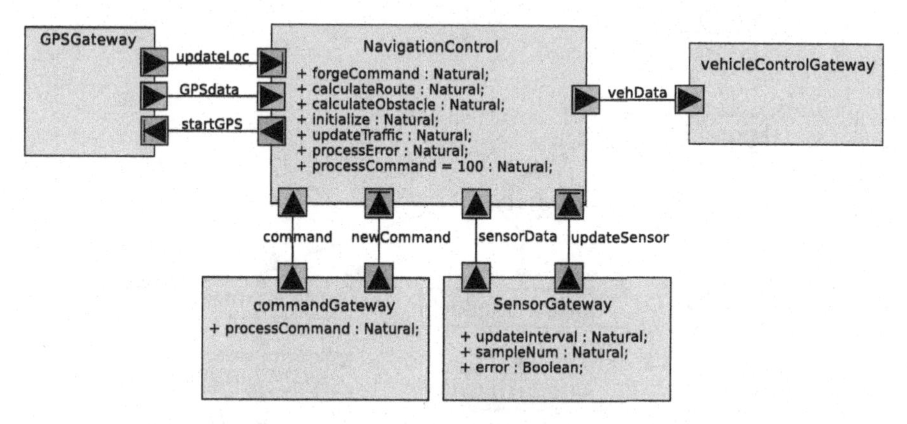

Figure 9.6. *Application view: component diagram of autonomous vehicle system. For a color version of the figure, see www.iste.co.uk/camara/wireless3.zip*

The activity diagram of the Navigation Task is shown in Figure 9.7. The navigation system starts with an initialization sequence modeled only as an execution time. Next, the *Send Event* operator signals a start to the GPS Gateway. After the initialization, the navigation task continually waits for any of the three possible signals from the other tasks as indicated by the *Select Event* operator. For example, if it receives the "update Location" event from the GPS, it acquires the new GPS data with the *Read Channel* operator. Next, it calculates the route based on the new data. If route calculation fails, it processes the error, and if it succeeds, it forges a new command and sends it to the Vehicle Control Gateway by writing data to the "vehData" channel. We model these possibilities with the *Choice* operator. The *loop forever* operator indicates that when one execution branch finishes, execution returns to the start of the loop, and executes the *Select Event* operator again.

9.4.3.2. *Architectural view*

The architectural model displays the underlying architecture as a network of abstract execution nodes, communication nodes and storage nodes. Execution nodes consist of CPUs and Hardware Accelerators, defined by parameters for simulation. All execution nodes must be described by data size, instruction execution time and clock ratio. CPUs can be further customized with specific parameters, e.g. cache miss percentage, and with information regarding operating system and middleware properties, e.g. scheduling policy, task switching time, etc.

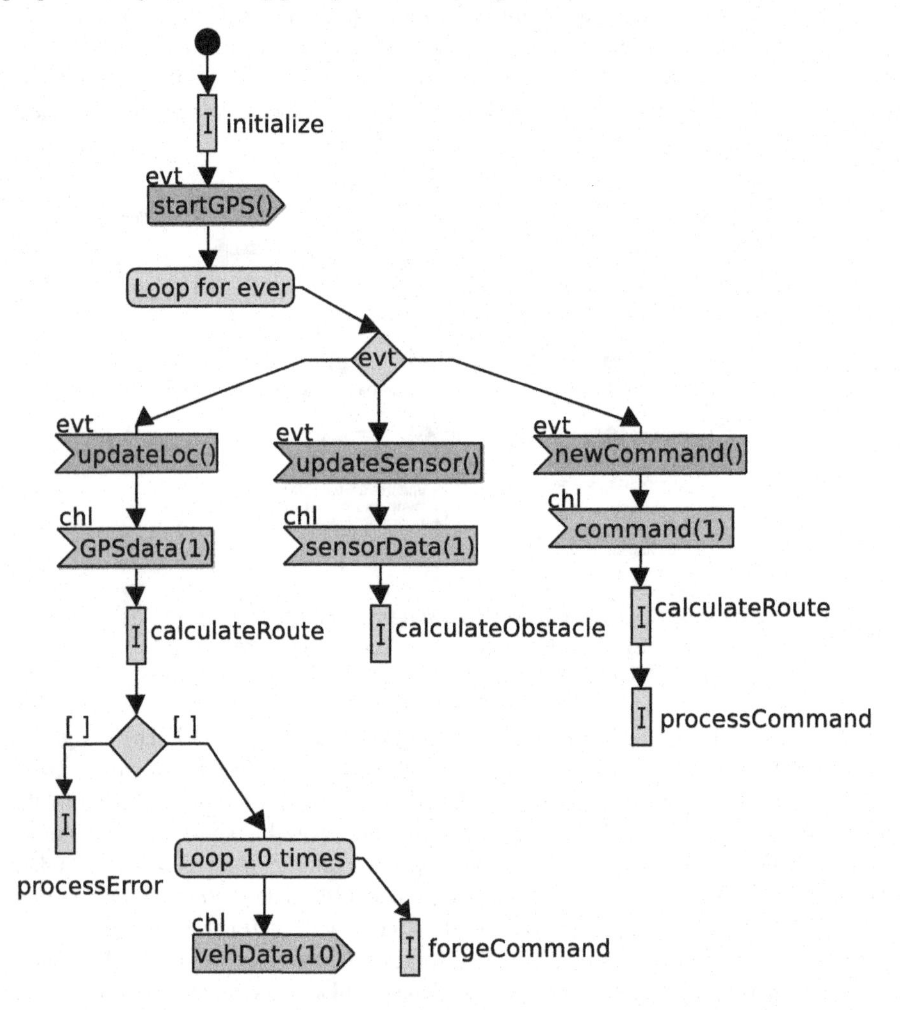

Figure 9.7. *Application view: activity diagram for autonomous vehicle navigation system*

Communication nodes include bridges and buses. Buses connect execution and storage nodes for task communication and data storage or exchange, and bridges connect buses. Buses are characterized by their arbitration policy, data size, clock ratio, etc., and bridges are characterized by data size and clock ratio. Storage nodes are Memories, which are defined by data size and clock ratio.

9.4.3.3. *Mapping view*

Mapping partitions the application into software and hardware as well as specifying the location of their implementation on the architectural model. A task mapped onto a processor will be implemented in software, and a task mapped onto a hardware accelerator will be implemented in hardware. The exact physical path of a data/event write may also be specified by mapping channels to buses and bridges or through associated Communication Patterns [ENR 14].

9.4.4. *Software/system design*

Finally, the *Software/System Design* stage (lower right of Figure 9.3) develops the functions mapped onto processors at the previous stage. A Partitioning model may be automatically translated into the Software/System Design model. Functions to be implemented are first analyzed with SysML-based use case and scenario views to determine a software design. A software design consists of SysML blocks and their interactions, such as operations on data, signal exchange and security protocols. Formal verification intends to prove correct functionality and resilience of the system under design to threats. The functional design is refined until it can be transformed to prototyping code.

Figure 9.8 shows the model of the autonomous system and the environment. As sensor and navigation algorithms are added, the system's response to various environmental conditions can be simulated and studied.

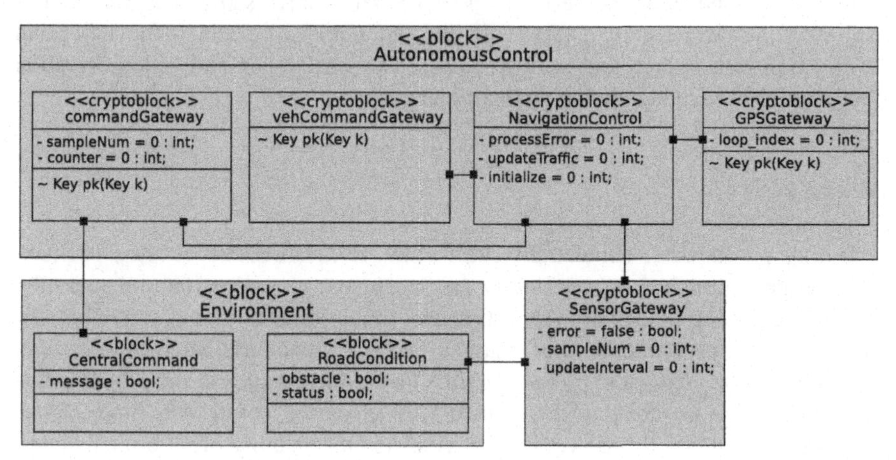

Figure 9.8. *Autonomous vehicle software/system design model*

9.4.5. *Assumptions*

Iterations over the complete method are assumption driven [DES 16]. More precisely, the system specification is first limited in scope, then progressively advanced to include further details. For example, we might first develop the design for functionality of the system, assuming no attacks are possible. In the next iteration, we remove the assumption on complete lack of attackers and add in security mechanisms, continually evolving the design.

Typical assumptions that we could take when designing autonomous cars are "the sensors will never fail" (assumption on the environment), "the power supply of the ECUs will never fail" (assumption on the environment) and "the plausibility check function will be ignored" (assumption on the system to be designed). Then, progressively, the system will be refined, and these assumptions will be removed, apart if they are part of the system specification.

9.4.6. *Tooling*

All modeling, formal verification, simulation and code generation are performed within the supporting toolkit TTool. TTool supports the automatic proof from diagrams at the push of a button. Safety proofs can be performed with integrated model checkers or with UPPAAL [BEN 04]. Security proofs are done with ProVerif [BLA 09]. For user convenience, results of verifications are also back traced to the graphical models, without having to investigate the results provided by UPPAAL and ProVerif.

Figure 9.9 shows a screenshot of TTool. The center shows Diagram Editor displaying the Architecture Diagram of the autonomous vehicle. The left panel shows navigation, search and analysis tools. The verification and code generation tools are displayed along the top toolbar as labeled.

9.4.7. *Safety*

During partitioning, simulations and formal verification can be used to study safety properties such as schedulability and worst case execution time on a specific architecture. Using execution cycle estimates of algorithms allows us to predict and simulate system performance. For example, in the Autonomous Navigation system, we ensure that the maximum processing time for obstacle detection should fall below a certain time to ensure that the vehicle will avoid the obstacle in time. Once mapping is complete, formal verification confirms reachability of all critical states. Simulations take into account time and show the step-by-step execution of the application, along with CPU and bus usage. For example, we may wish to determine if each sample of sensor data is processed by the Navigation Unit within a certain time, or ensure that the vehicle will respond in time to dodge an obstacle. High bus or CPU contention may encourage us to modify the architecture or redo the mapping, such as providing each Gateway unit with its own CPU.

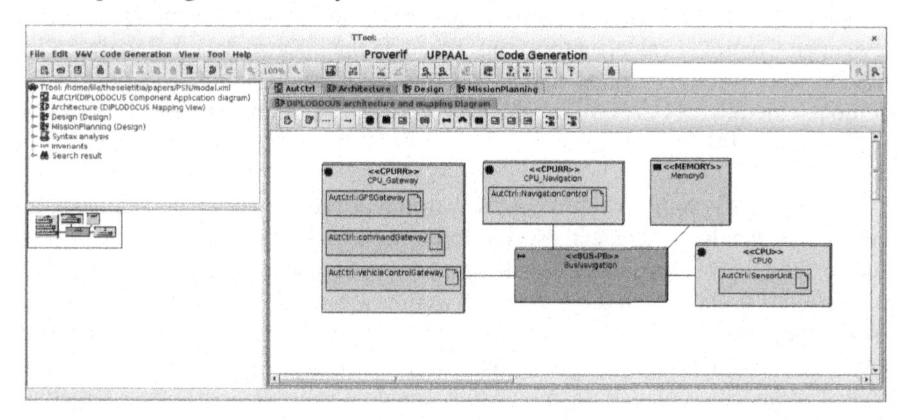

Figure 9.9. *Screenshot of TTool*

UPPAAL is a model checker for networks of timed automata. The behavioral model of a system to be verified is translated into a UPPAAL specification. Safety Proofs conducted with UPPAAL check for unwanted behavior through translation of the model into automata. For example, UPPAAL may verify the lack of deadlock, such as two threads both waiting for the other to send a message. Behavior may be verified through “Reachability”, “Leads to” and other general statements. “Leads to” allows us to verify that one state must always be followed by another. For example, in the Autonomous Response Vehicle model, “Abort Mission” by the central command should always be followed by Navigation control executing “Return to base”. The model checker can also verify a statement for all execution flows, A[], or the existence of a single execution flow E<>. For example, it may verify that a state “Route Found” is reachable with E<> NavigationControl.RouteFound. Critical properties must be preserved at all times, such as the vehicle not continuing to drive when the road is blocked. We might also verify that the Autonomous Vehicle does not send the message when a process error is detected.

9.4.8. *Security*

In our context (security in embedded systems), we focus on the properties of authenticity and confidentiality. These properties are explained in the scope of rescue missions with UAVs and autonomous cars.

Authenticity (or integrity) confirms that a piece of data received by an entity really did originate from the expected sender, and that it was not tampered with. It means that, for a particular exchange, a received message really does correspond to a message sent by the expected entity. In an embedded system, a failure to enforce authenticity may allow an attacker to forge messages, impersonate a trusted component (such as a remote controller) and change the behavior of a system. For example, an attacker forging sensor data may cause an autonomous vehicle's Navigation Control to calculate trajectory based on the injected data, fail to detect an obstacle, and end up in an accident, resulting in mission failure, and potentially additional victims. UAVs have already been shown to be vulnerable to GPS spoofing [KER 14].

Confidentiality, in our context, deals with the privacy of sensitive data, such as personal information or credentials. For instance, keys stored in a Trusted Platform Module should probably not be broadcast across an insecure channel. For proposed disaster-relief drones, captured images of victims are sensitive and, for the sake of privacy, they should not be allowed to be intercepted and posted online or published [TAN 15b]. Also, drones capable of generating a 3D mapping of a building carry a detailed architectural plan of an area, which could be valuable to criminals targeting this location. Even if an attacker physically steals a drone, the on-board sensitive information should remain confidential.

Security analysis is performed with ProVerif, a verification tool operating on designs described in pi-calculus [BLA 09]. A ProVerif specification consists of a set of processes communicating on public and private channels. Processes can split to create concurrently executing processes, and replicate to model multiple executions (sessions) of a given protocol. Cryptographic primitives such as symmetric and asymmetric encryption or hash can be modeled through constructor and destructor functions. ProVerif assumes a Dolev-Yao attacker, which is a threat model in which anyone can read or write on any public channel, create new messages or apply known primitives.

ProVerif provides its user with the capabilities to query the confidentiality of a piece of data, the authenticity of an exchange or the reachability of a state. Traces are generated for all possible execution paths. The tool then presents a result to the user that is either *true* if the property is verified, *false* if a trace that falsifies the property has been found, or *cannot be proved* if ProVerif failed in asserting or refuting the queried property.

9.4.9. *Security modeling*

Security properties can be modeled starting from the Partitioning Phase. On the architectural modeling, buses can be specified as public or private. For example, devices communicating on a WiFi network would be modeled as exchanging over a public bus, while the internal bus would be modeled as private. Private buses are marked *secure* with a green shield. The distinctions between bus types also model assumed attacker capabilities: if we assume that an attacker has no physical access to the system, then we can describe internal buses as private, but if an attacker could physically probe the bus, then it must be indicated as public.

On the application modeling, we use *Cryptographic Configurations* to model security elements. The configuration types include Symmetric Encryption, Asymmetric Encryption, Hash, Nonce, Key, etc. For each security type, the configuration provides by default estimated values for overheads (additional bits added to the message) and Computational Complexity (additional execution cycles) to help engineers less familiar with security mechanisms. A *Cryptographic Configuration* can be added as a tag to channels to indicate the data exchanged was first encrypted before transfer. Nonces can be added to messages before encryption.

During the Software/System Design phase, crypto-blocks and encryption/decryption operators are used to model how data is secured. The exchange and subsequent encryption of data is mathematically analyzed to determine if it is vulnerable to an attacker. For example, the attacker should not be able to forge messages between the command center and autonomous emergency vehicle. Also, we wish to verify the strong authenticity of a message, protecting against replay attacks, i.e. an attacker recoding a previously sent message, and sending it again later on.

9.5. Mission planning

This section investigates how to formally validate that within a given mission, the system under design in our case, the autonomous system will correctly handle its mission in a safe, secure and efficient way.

It is unrealistic to test all possible outcomes and execution flows of a mission by hand, and logical flaws or undesired corner cases may not be obvious in a larger model. In particular, the partitioning and design of the system is evaluated with regards to a given environment model, and for a set of standard missions and situations. Thus, for a specific rescue mission, we suggest modeling the mission and then formally verifying it with TTool. Combined with a model of the selected equipment, we can ensure that the mission is possible before sending out the vehicles. We can add “Error states” within the activity diagram, where the state cannot be reached in a normal operation, and then use UPPAAL to verify the non-reachability of that state.

The methodology is shown in Figure 9.10. We start by providing a general behavioral model of the vehicles to be used, which is then refined until it is successfully validated. Then, the specific mission itself is modeled. The formal verifier determines if the mission is feasible under given constraints. If verification of the mission fails, then the engineer must either revise the mission parameters or choose different vehicles to perform the mission, and perform verification again. If the mission is verified possible, then TTool generates a code automatically to be loaded aboard the vehicle. For example, we can determine if the UAV is capable of reaching a destination in a given time. We present a simple example with one UAV and one autonomous vehicle.

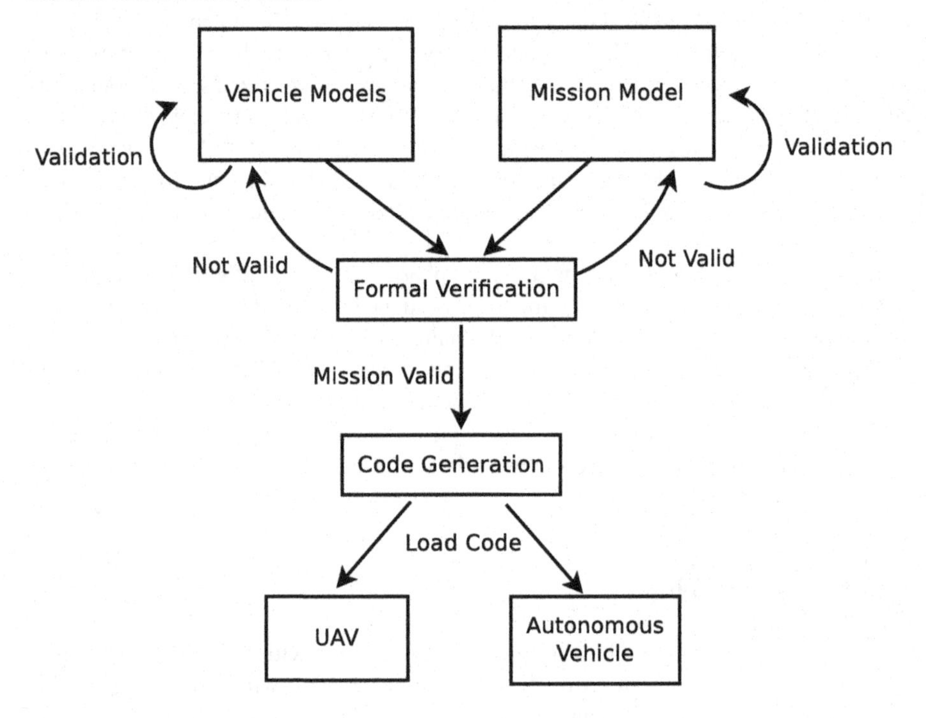

Figure 9.10. *Mission planning flowchart*

While it is unnecessary to ensure that the mission succeeds in all possible circumstances, it is necessary to verify that the mission is capable of succeeding. However, it is necessary to ensure that no unsafe situations occur, such as the autonomous vehicle continuing to drive on a road which has been detected damaged. In a mission to deploy a mobile relay, typical mission verification would involve estimating the feasibility of the solution, taking into account battery life, travel distance and length of deployment.

Mission planning can involve several vehicles at the same time, and investigate how/when they can collaborate to better fulfill the mission. For example, in the case of a mobile relay, the mission planner can be used to determine how many UAVs shall be used to provide a relay in a given location, taking into account the time to reach the position, the operational time, the time to return and finally the time to recharge the batteries. It could also be used to model how many backup drones shall be used, and where they should be positioned to minimize network failure whenever a current UAV relay fails.

9.5.1. *Mission description*

In this example, we propose a mission with 1 UAV, 1 Autonomous Emergency Vehicle and Central Command which coordinates with both objects. The UAV must scout out a set of locations and send back images to determine if roads are accessible to the vehicle. Based on road conditions, Central Command directs the Delivery Vehicle to destinations. The Autonomous Emergency Vehicle delivers supplies to directed destinations until none remains, and then returns to the base.

The detailed activity diagrams for the actors are shown in Figures 9.11 and 9.12. Central Command continually monitors the status of both vehicles, and if inclement weather arises, it sends an abort mission message to direct them to return. Using the image data sent from the UAV, Central Command automatically processes the data to detect whether a destination is reachable or not for the Emergency Vehicle, and sends the Emergency Vehicle to the new destination. If no delivery destinations remain, then it directs the Emergency Vehicle to return to the base. Red Xs mark states that UPPAAL should verify reachability.

Many vital safety properties must be verified for the mission. For example, the UAV must verify roads to a destination are accessible before the Emergency Vehicle begins to travel. Therefore, there must be no state where the Emergency Vehicle is traveling while the UAV has returned to base. To query this statement in UPPAAL, we write

```
"E<> UAV.return && AV.travel"
```

The statement verifies if there exists a state where the UAV is in the return state and the Emergency Vehicle is still traveling.

We also verify correctness in behavior. For example, if Central Command directs to abort the mission, then both the UAV and Emergency Vehicle will return. We express this with a "Leads to" statement,

```
"CentralCommand.abortMission --> UAV.return"
```

In other words, every time state "abortMission" is reached, it must eventually be followed by the state "return".

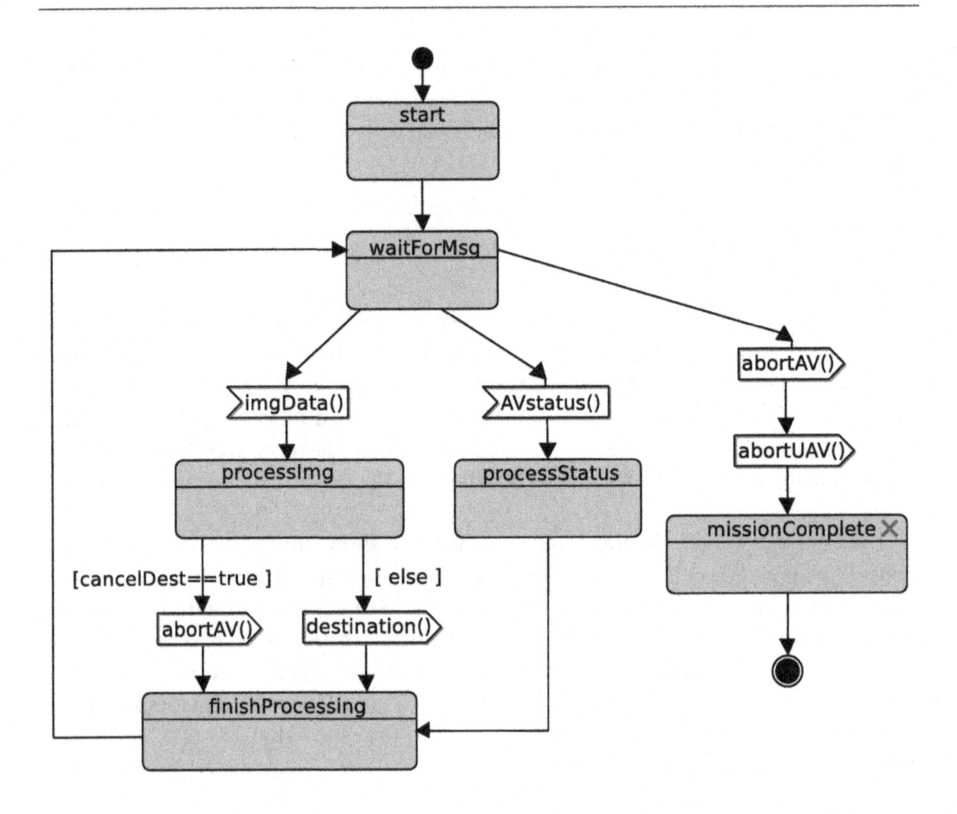

Figure 9.11. *Central Command activity diagram*

These statements may be entered directly on the UPPAAL model checker, or permanently stored on the model as pragma to be verified in UPPAAL. Their most recent verification status is displayed as an X or a check mark. The pragma is displayed on the block diagram as shown in Figure 9.13. As shown, UPPAAL has verified that both the UAV and Emergency Vehicle will return after an "abort mission" command. Also, there is never the case where the Emergency Vehicle is traveling while the UAV has returned.

9.5.2. *Integration of mission planning and autonomous vehicles*

In Figure 9.14, we evaluated the partitioning of the AV alone. The figure shows the mapping view back traced with the simulation information, e.g. the CPU and bus load. Terminated tasks are shown in red and blocked tasks in orange.

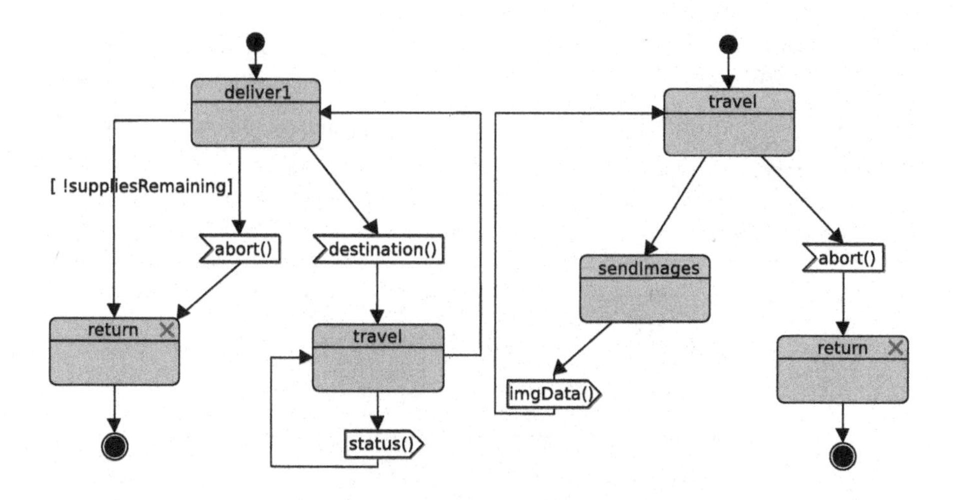

Figure 9.12. *Autonomous delivery vehicle and unmanned aerial vehicle activity diagrams*

We add the mission-specific details to the command module to evaluate the performance impact in order to determine if the partitioning of the system can still be successfully validated. With the mission loaded, the task Navigation Control must process the new commands, increasing the load on its CPU as shown in Figure 9.15. Other CPUs and the bus have a lower load because the task "Navigation Control" now creates starvation for other tasks. Thus, a better partitioning would be to increase the CPU frequency, or to discharge this task with hardware accelerators.

9.6. Related work

9.6.1. *Embedded system design*

Many toolkits support modeling of embedded systems. They support various stages of the design process.

Capella [POL 08] relies on Arcadia, a comprehensive model-based engineering method. It provides architecture diagrams allocating functions to components, allocation of Behavioral Components onto Implementation Components (typically hardware, but not necessarily). Capella provides advanced mechanisms to model bit-precise data structures, and relate them to Functional Exchanges, Component or Function Ports, Interfaces, etc. Capella is, however, more business focused and supports multiple methodologies.

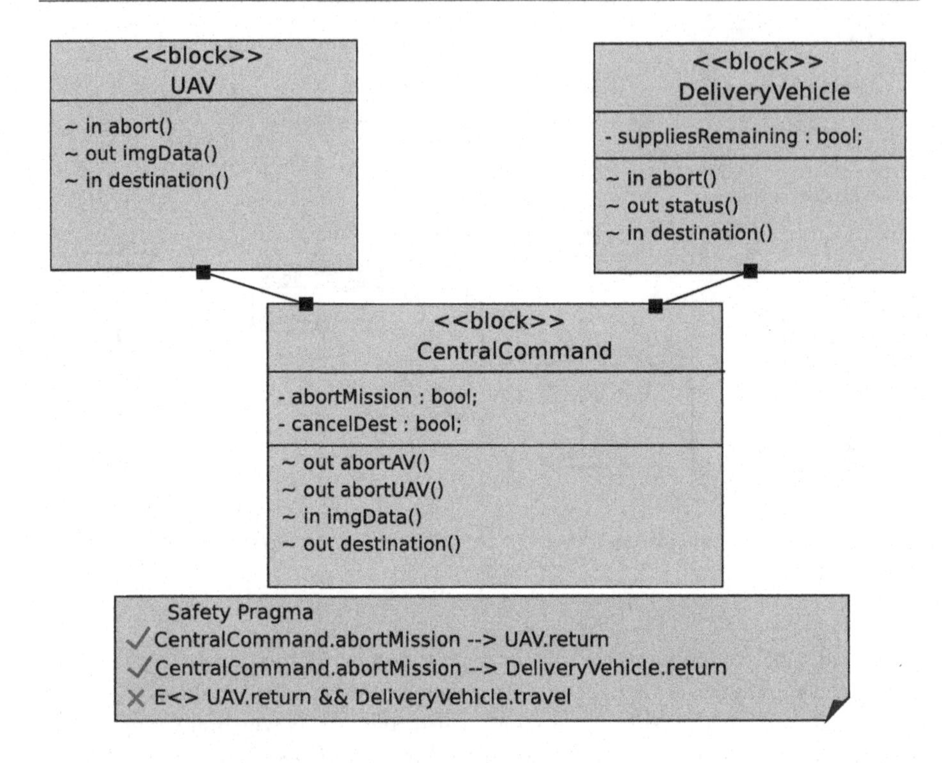

Figure 9.13. *Mission model showing verified safety properties*

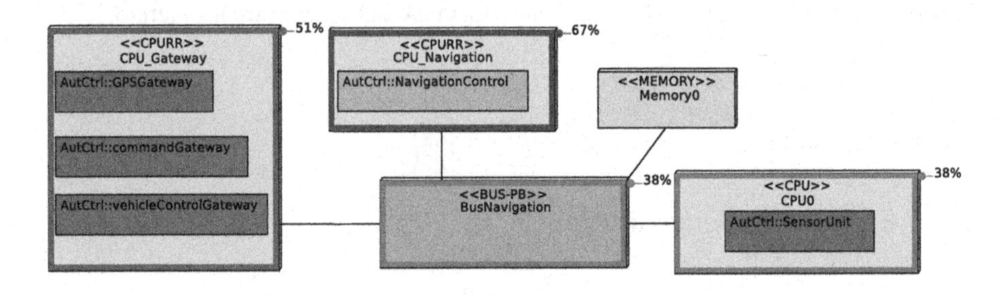

Figure 9.14. *Simulation for system without mission. For a color version of the figure, see www.iste.co.uk/camara/wireless3.zip*

Design Space Exploration (DSE) of Systems-on-Chip is the process of analyzing various functionally equivalent implementation alternatives to select an optimal solution [MUH 06]. The most suitable design is commonly chosen based on metrics such as *functionality*, *performance*, *cost*, *power*, *reliability* and *flexibility*. At the system level, DSE is challenging because the system design space is extremely large and so usual simulation-based analysis techniques fail to efficiently observe the above mentioned metrics. Contributions to DSE environments such as [BAL 03, WOL 99, CHA 01, ASS 07, SCH 05, KUK 05, VID 09] generally rely on a high-level language to describe application functions and architectures. For example, [KUK 05, SCH 05, VID 09] rely on UML or MARTE diagrams.

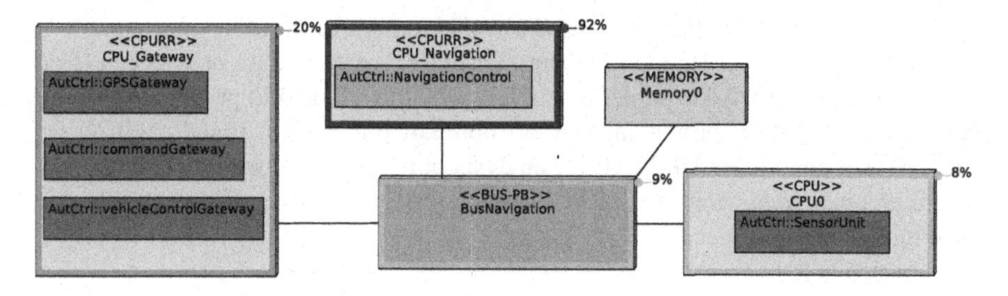

Figure 9.15. *Simulation for system with mission*

The Architecture Analysis & Design Language (AADL [FEI 04]) is a standard from the International Society of Automotive Engineers (SAE). It allows the use of formal methods for safety-critical real-time systems in avionics, automotives and other domains. Similar to our environment, a processor model can have different underlying implementations and its characteristics can easily be changed at the modeling stage.

9.6.2. *Safety*

As safety is important in the management of UAVs and other robotic systems, many previous works have used formal methods to validate mission planning or autonomous systems. [KVA 10] used a hybrid partial order forward chaining framework for mission planning. The framework monitors the states of its agents and formally determines allowed new actions based on the current state structure. The system is intended to be used for collaborative Unmanned Aircraft Systems. Similarly, [GJO 12] formally validated collective robot systems with KLAIM. Their approach studies robots with a common goal, communicating with each other instead of a central command. Verification analyzes the probabilities of fulfillment of desired formal properties expressed in stochastic logic MoSL. Our approach relies on central command to communicate with individual autonomous agents and dynamically alter the mission in accordance with environmental conditions. Our verification in UPPAAL relies on states and reachability properties to determine mission success.

Other works validated the architecture and design of the autonomous systems themselves. [WEB 11] used Promela/Spin and Gwendolen/AJPF for rule checking Unmanned Aircraft Systems. The UAS must obey "Rules of the Air", such as detecting collisions and evading them. SPIN/PROMELA were used for fast, high-level modeling, and Gwendolen for slower, detailed modeling of interactions. [CHA 10] validated the safety architecture of autonomous systems with

Event-B. The work intended to verify that the probability of a crash would be less than 10^{-9} per flight hour, and a single failure did not cause a crash. Our safety verification, however, intends to verify system behaviors and is not concerned with probabilities.

Generally, UPPAAL has been used for the formal verification of protocols or designs, such as the Ad Hoc On-Demand Distance Vector routing protocol in [FEH 12]. The derived UPPAAL model of the protocol was verified to process all routing messages, and after processing, only optimal routes are found. [HEN 06] demonstrated how to use UPPAAL for analysis of timeliness properties of embedded system architectures. However, their automata were manually generated while ours are automatically generated.

9.6.3. *Security*

Many projects verify security properties of embedded systems and security protocols. [JAV 12] presents a threat model of possible attacks on UAVs to assist designers with risk analysis. Attacks are categorized into Confidentiality, Integrity or Availability, and then further separated by origin or location of attack.

The Knowledge Acquisition in Automated Specifications approach Security Extension aims to identify security requirements for software systems [LAM 04]. The methodology uses a goal-oriented framework and builds a model of the system, and then an anti-model which describes possible attacks on the system. Both models are incrementally developed: threat trees are derived from the anti-model and the system model adds security countermeasures to protect against the attacks described in the anti-model.

The Combined Harm Assessment of Safety and Security for Information Systems (CHASSIS) method considers safety and security together in a common model [RAS 13]. Safety and security hazards in the form of misuse cases are developed, and then trade-off analysis is used to unify all requirements and identify when safety and security conflict. While these techniques targeting the requirements and analysis phase offer a detailed approach to considering threats against safety and security, they are not yet automated.

Other approaches offer comprehensive modeling of security mechanisms intended for the Software/System Design phase. For example, UMLSec [JÜR 02] is a UML profile for expressing security concepts, such as encryption mechanisms and attack scenarios. It provides a modeling framework to define security properties of software components and of their composition within a UML framework. It also features a rather complete framework addressing various stages of model-driven secure software engineering from the specification of security requirements to tests, including logic-based formal verification regarding the composition of software components.

SecureUML enabled the design and analysis of secure systems by adding mechanisms to model role-based access control [LOD 02]. Authorization constraints are expressed in Object Constraint Language (OCL) for formal verification. Our security model focuses on protecting against an external attacker instead of access control. In contrast to formula-based constraints or queries, our approach to security analysis relies on graphically annotating the security properties to query within the model. Our methodology considers security at all stages of the design process, validating partitioning models as well as protocols.

9.7. Conclusion and perspectives

UAVs and autonomous vehicles may perform routine or dangerous tasks, assisting rescue workers during disaster relief efforts. However, it is important to ensure these autonomous objects are safe and secure, so that they may not be hijacked, leak sensitive data, or further injure victims. We presented how the toolkit TTool can automatically verify missions or autonomous vehicles, requiring no knowledge of the verification languages for a designer.

The main advantage of these UAVs and autonomous vehicles is their ability to free up relief workers to focus on critical tasks. Mission planning and management should require as little manual input as possible. However, mission planning with autonomous objects requires specification of a mission, vehicle characteristics and communications between central command and all vehicles. Currently, TTool performs validation and code generation automatically, but still requires a user to manually enter the mission diagrams.

In future work, we consider the addition of new views/diagrams specific to mission planning. Libraries/Patterns modeling common situations/mission requirements will allow for easier and more efficient design of missions. Instead of manually building states and transitions in the diagram, these mission patterns could allow a user to specify only the high-level behavior and automatically receive the low-level implementation details. For example, a user might specify only "deploy at location X for time Y", and the UAV would generate their activity diagram, including navigation, battery management, etc.

Furthermore, we will increase the automatic exploration capabilities of TTool, such as automatic selection of vehicles capable of carrying out the current mission. Generic models for common autonomous objects could be provided, allowing the user to customize them with only certain device-specific parameters such as speed, battery life, etc. These mission-focused additions to TTool will better assist relief workers with the automatic planning and management of UAVs and autonomous vehicles, improving the execution and efficiency of rescue efforts during disasters.

9.8. Acknowledgment

The authors would like to thank their sponsor, Vedecom, an institute for the development of electric, connected and autonomous vehicles.

9.9. Bibliography

[APV 15] APVRILLE L., ROUDIER Y., "SysML-Sec attack graphs: compact representations for complex attacks", *Second International Workshop on Graphical Models for Security (GraMSec 2015)*, Verona, Italy, Springer, LNCS, vol. 9390, pp. 35–49, July 2015.

[ASS 07] ASSAYAD I., YOVINE S., "Performance analysis of embedded multiprocessor industrial applications: methodology and tools", *14th IEEE International Conference on Electronics, Circuits and Systems*, Marrakech, Morocco, pp. 907–910, June 2007.

[BAL 03] BALARIN F., *Hardware-Software Co-Design of Embedded Systems, The POLIS Approach*, 5th edition, Kluwer Academic Publishers, 2003.

[BEN 04] BENGTSSON J., YI. W., "Timed automata: semantics, algorithms and tools", REISIG W., ROZENBERG G. (eds.), *Lecture Notes on Concurrency and Petri Nets*, LNCS 3098, Springer-Verlag, pp. 87–124, 2004.

[BLA 09] BLANCHET B., "Automatic verification of correspondences for security protocols", *Journal of Computer Security*, vol. 17, no. 4, pp. 363–434, July 2009.

[CHA 01] CHATELAIN A., MATHYS Y., PLACIDO G. *et al.*, "High-level architectural co-simulation using Esterel and C", *Ninth International Symposium on Hardware/Software Codesign. CODES 2001 (IEEE Cat. No.01TH8571)*, pp. 189–194, April 2001.

[CHA 10] CHAUDEMAR J.-C., BENSANA E., SEGUIN C., "Model based safety analysis for an unmanned aerial system", *DRHE 2010 – Dependable Robots in Human Environments*, Toulouse, France, 2010.

[CHE 11] CHECKOWAY S., MCCOY D., KANTOR B. *et al.*, "Comprehensive experimental analyses of automotive attack surfaces", *USENIX Security Symposium*, San Francisco, 2011.

[DES 16] DE SAQUI-SANNES P., APVRILLE L., "Making modeling assumptions an explicit part of real-time systems models", *8th European Congress on Embedded Real Time Software and Systems (ERTS2'2016)*, Toulouse, France, January 2016.

[ENR 14] ENRICI A., APVRILLE L., PACALET R., "Model-Driven Engineering Languages and Systems", *17th International Conference, MODELS 2014*, Valencia, Spain, 28 September–3 October, 2014.

[FEH 12] FEHNKER A., VAN GLABBEEK R., HÖFNER P. *et al.*, *Automated Analysis of AODV Using UPPAAL*, Springer, Berlin-Heidelberg, 2012.

[FEI 04] FEILER P.H., LEWIS B.A., VESTAL S. *et al.*, "An overview of the SAE architecture analysis & design language (AADL) standard: A basis for model-based architecture-driven embedded systems engineering", in DISSAUX P., FILALI-AMINE M., MICHEL P. *et al.* (eds.), *IFIP-WADL*, Springer, 2004.

[GJO 12] GJONDREKAJ E., LORETI M., PUGLIESE R. *et al.*, *Towards a Formal Verification Methodology for Collective Robotic Systems*, Springer, Berlin-Heidelberg, 2012.

[HEN 06] HENDRIKS M., VERHOEF M., "Timed automata based analysis of embedded system architectures", *Proceedings 20th IEEE International Parallel Distributed Processing Symposium*, p. 8, April 2006.

[JAV 12] JAVAID A.Y., SUN W., DEVABHAKTUNI V.K. *et al.*, "Cyber security threat analysis and modeling of an unmanned aerial vehicle system", *2012 IEEE Conference on Technologies for Homeland Security (HST)*, pp. 585–590, November 2012.

[JÜR 02] JÜRJENS J., "UMLsec: extending UML for secure systems development", *Proceedings of the 5th International Conference on the Unified Modeling Language*, London, UK, pp. 412–425, 2002.

[KER 14] KERNS A.J., SHEPARD D.P., BHATTI J.A. *et al.*, "Unmanned aircraft capture and control via GPS spoofing", *Journal of Field Robotics*, vol. 31, no. 4, pp. 617–636, Wiley Online Library, 2014.

[KIE 02] KIENHUIS B., DEPRETTERE E., VAN DER WOLF P. *et al.*, "A methodology to design programmable embedded systems: the Y-chart approach", *Embedded Processor Design Challenges*, Springer, pp. 18–37, 2002.

[KOS 10] KOSCHER K., CZESKIS A., ROESNER F. *et al.*, "Experimental security analysis of a modern automobile", *2010 IEEE Symposium on Security and Privacy*, IEEE, pp. 447–462, 2010.

[KUK 05] KUKKALA P. *et al.*, "Performance modeling and reporting for the UML 2.0 design of embedded systems", *Proceedings of the 2005 International Symposium on System-on-Chip*, pp. 50–53, November 2005.

[KVA 10] KVARNSTRÖM J., DOHERTY P., "Automated planning for collaborative UAV systems", *2010 11th International Conference on Control Automation Robotics Vision (ICARCV)*, pp. 1078–1085, December 2010.

[LAM 04] VAN LAMSWEERDE A., "Elaborating security requirements by construction of intentional anti-models", *Proceedings of the 26th International Conference on Software Engineering, ICSE '04*, pp. 148–157, 2004.

[LOD 02] LODDERSTEDT T., BASIN D., DOSER J., "Secure UML: a UML-based modeling language for model driven security", *5th International Conference on the Unified Modeling Language*, pp. 426–441, 2002.

[MIL 15] MILLER C., VALASEK C., "Remote exploitation of an unaltered passenger vehicle", *Black Hat USA*, 2015.

[MUH 06] MUHAMMAD W., LUDOVIC A., RABEA A.-B. *et al.*, "Abstract Application Modeling for System Design Space Exploration", pp. 331–337, 2006.

[POL 08] POLARSYS, "ARCADIA/CAPELLA", available at: https://www.polarsys.org/capella/arcadia.html, 2008.

[RAN 13] RANFT B., DUGELAY J.-L., APVRILLE L., "3D perception for autonomous navigation of a low-cost MAV using minimal landmarks", *International Micro Air Vehicle Conference and Flight Competition*, Toulouse, France, September 2013.

[RAS 13] RASPOTNIG C., KATTA V., KARPATI P. *et al.*, "Enhancing CHASSIS: a method for combining safety and security", *Eighth International Conference on Availability, Reliability and Security (ARES)*, pp. 766–773, September 2013.

[ROU 13] ROUDIER Y., IDREES M., APVRILLE L., "Towards the model-driven engineering of security requirements for embedded systems", *Model-Driven Requirements Engineering Workshop (MoDRE)*, Rio de Janeiro, Brazil, July 2013.

[SCH 05] SCHATTKOWSKY T. *et al.*, "A model-based approach for executable specifications on recon figurable hardware", *Design, Automation and Test in Europe Conference and Exhibition, 2005. DATE'05*, pp. 692–697, November 2005.

[TAN 14] TANZI T., APVRILLE L., DUGELAY J.-L. *et al.*, "UAVs for humanitarian missions: autonomy and reliability", *IEEE Global Humanitarian Technology Conference (GHTC)*, Silicon Valley, USA, October 2014.

[TAN 15a] TANZI T., ISNARD J., "Public Safety Network: An Overview" in CÂMARA D., NIKAEIN N. (eds.), *Wireless Public Safety Networks 1*, ISTE Press, London and Elsevier, Oxford, 2015.

[TAN 15b] TANZI T., SEBASTIEN O., RIZZA C., "Designing autonomous crawling equipment to detect personal connected devices and support rescue operations: technical and societal concerns", *The Radio Science Bulletin*, vol. 355, no. 355, pp. 35–44, International Union of Radio Science (U.R.S.I.), 2015.

[VID 09] VIDAL J., DE LAMOTTE F., GOGNIAT G. *et al.*, "A co-design approach for embedded system modeling and code generation with UML and MARTE", *Design, Automation and Test in Europe Conference and Exhibition, 2009. DATE'09*, pp. 226–231, April 2009.

[WEB 11] WEBSTER M., FISHER M., CAMERON N. *et al.*, *Formal Methods for the Certification of Autonomous Unmanned Aircraft Systems*, Springer, Berlin-Heidelberg, 2011.

[WOL 99] WOLF P.V.D. *et al.*, "A methodology for architecture exploration of heterogeneous signal processing systems", *IEEE Workshop on Signal Processing Systems (SiPS99)*, 1999.

10

Disaster Resilient Telematics Based on Device-to-Device Communication

Infrastructure-less communication networks, like ad-hoc networks, are more resilient and can maintain communications during disaster events. However, supporting critical telematic services during and in the aftermath of a disaster using ad-hoc networks has proven to practically be a non-trivial task primarily due to the various issues faced when trying to apply Internet protocols to the wireless environment. In response, research and development recently has been steered towards new beaconing-based ad-hoc networking approaches that are decentralized, distributed and direct in nature. These approaches ensure communication resilience thanks to the opportunistic nature of beaconing-based neighborhood discovery and communication. At the same time, they adapt faster to local changes in their vicinity (an action that matters most to distress survivors). They also ensure network scalability since each device communicates only with its direct neighbors and data forwarding occurs in a hop-by-hop manner. This chapter reviews current research and standardization efforts for beaconing-based ad-hoc

Chapter written by Panayiotis KOLIOS, Christos PANAYIOTOU and Georgios ELLINAS.

communications. Subsequently, the networking aspects that arise from beaconing-based communication are documented and discussed. Thereafter, critical telematic applications and services tailor-made to this type of networking approach are identified, and for a subset of those, novel solutions are presented.

10.1. Introduction

Similarly to any other utility infrastructure, the extent of damage to a telecommunications network greatly depends on the strength by which a particular area is hit during a natural or man-made disaster. In the case of Hurricane Katrina [FCC 06] in the USA (August 2005), 3 million landlines and at least 1000 cell towers were disabled. During the Haiti earthquake [COR 10] (January 2010), all cellular networks were severely affected to the extent that they became non-operational, either due to the infrastructure damage around Port-au-Prince or due to severe congestion. The effects of the Tohoku earthquake in Japan (March 2011) on the telecommunications infrastructure were equally dramatic. In total, 1.9 million communication lines and about 29,000 cellular base stations suffered extensive damage. For the remaining (operational) networks, carriers were forced to restrict voice traffic by 70–95%. During that emergency, NTT Docomo had to suspend operations within the Tohoku region from the first day and was only able to recover services to an almost pre-disaster condition more than a month later (26 April 2011) [NTT 12]. During Hurricane Sandy (October 2012), an estimated 25% of the total base stations were damaged over a vast geographical area of the USA east coast, due to the unprecedented large size of the storm surge. Those high storm surge levels caused extensive flooding that in turn triggered power outages which were the main cause of network unavailability. The latter sequence of events also demonstrated the increasing interdependencies between utility infrastructures that hinder proper network planning [KWA 12].

Undoubtedly, communication and information exchange are vital in these situations. For one, communication enables social interactions and thus mediates reassurance between stranded survivors. Furthermore, it also facilitates exchange the information of between stranded survivors and emergency responders, improving situational awareness and subsequently allowing first responders to perform better triage and better planning for their search-and-rescue missions. Infrastructure-based solutions, however, are unreliable since they suffer from the catastrophic consequences of (natural or man-made) disasters [MAS 11], [GEO 10] and [CHE 10]. On the contrary, wireless ad-hoc networks opportunistically formed by mobile devices are far more resilient during disasters since they are physically detached from any infrastructure components. More importantly, ad-hoc networks can support decentralized, distributed and direct operations by default; making them robust to dynamic changes and failures [SUZ 12]. Excessive signaling associated with centralized solutions is also avoided when interactions only take place locally. At the same time, this approach increasingly becomes a feasible solution between consumer electronic devices with a plethora of new applications and services currently emerging [LIU 13].

Indicatively, there are already as many mobile subscriptions as people around the world [ITU 13] and the majority of these subscriptions are associated with full-fledged mobile computing devices (including basic handsets, smartphones and tablets), with the latest communication circuitry capable of receiving, processing and transmitting information both locally and within the infrastructure. And while it is well known that the battery life of these devices lasts for a couple of hours under normal use, it can adequately support basic telematic services for extended periods of time when processing and communication is thriftily used.

Up until recently, interconnecting devices in ad-hoc mode had been supported by the same Internet protocols used in traditional wireline communications. A review of the most prominent projects for emergency response networking that are based on the latter approach is presented in the sequel. As of late though, a new beaconing-based approach has emerged that has the potential to achieve a solution which is highly efficient, completely decentralized, distributed and direct communication. In a nutshell, this beaconing approach allows devices to duty cycle between the active and sleep communication states. During the sleep mode, devices conserve energy by shutting off their communication circuitry, while in the active state devices beacon their presence and listen for data. The beaconing mechanisms ensure proximate discovery and at the same time enable communications between neighboring devices. In fact, this approach has become so popular that all major wireless communication technologies are currently in the process of standardizing beaconing protocols to be integrated into their upcoming releases (including LTE-Direct, WiFi-Direct and Bluetooth Smart (Bluetooth Low Energy)). Importantly, these protocols are only concerned with the setup of a link between two devices and the proper exchange of data for the particular pair of devices. They are not designed to support internetworking, which is left to the designers of proprietary networking solutions.

In addition to reviewing current state-of-the-art ad-hoc networking solutions, this chapter presents a complementary ad-hoc networking architecture that is beaconing compliant and purposefully designed to address the constraints and requirements of ad-hoc public safety networks (PSNs). Ad-hoc PSNs are defined as networks that can be set up opportunistically between mobile devices during and after a disaster. They are formed by handheld devices to support the exchange of information among stranded survivors and between these survivors and emergency response units. The peculiarity of ad-hoc PSNs as opposed to traditional ad-hoc networks is that the source–destination pair for routing information is unknown. For instance, a stranded survivor in distress does not know who to contact to get help. Even worse, survivors do not know when emergency responders will arrive and what search-and-rescue procedure to follow. At the same time, a first responder does not know which survivors are in urgent need this assistance of scenario is shown in Figure 10.1. Even if a survivor's device could reach every other networked device, it would not know where to route its alert notifications since none of them have information on when and from which direction emergency responders will arrive. A stranded survivor can only communicate with those devices that are active at a particular instance in time and the emergency responders can only query active devices that are found within their search-and-rescue paths.

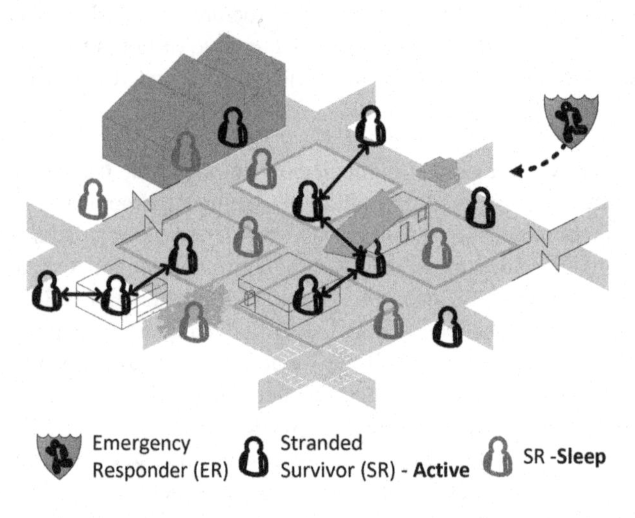

Figure 10.1. *Ad-hoc public safety network setup*

In that respect, key requirements of these ad-hoc PSNs include the survivability of the network for extended time periods, supporting the dissemination of alert messages across the network and supporting the preservation of alert messages even in the failure of individual devices.

10.2. Public safety ad-hoc networking

To date, viable commercial solutions for ad-hoc networking are non-existent. Instead, a number of community projects are being developed by academic institutions, public safety entities, non-profit organizations and individuals. As such, these projects are based mostly on open-source software that makes use of existing information and communication technologies to enable device-to-device communication. Table 10.1 provides a list of several projects that have recently received publicity due to their significant impact on disaster management.

Project name	Developers	Test sites	Technology enablers
Freifunk [FRE 16]	Grassroots initiative	Germany, Afghanistan, Ghana, Vietnam, etc.	WiFi mesh network
Village Telco [ADE 11]	Shuttleworth Foundation, in collaboration with Dabba	South Africa, East Timor, Brazil, Puerto Rico, Colombia, Nigeria	WiFi mesh network with analogue telephone adaptors
Commotion Wireless [GER 14]	Open Technology Institute	USA Communities including Washington, Brooklyn and Detroit	WiFi mesh network
Serval [GAR 13]	Resilient Networks Lab	Australia, USA, Nigeria, New Zealand, South Africa	WiFi ad-hoc and mesh network together with other long-range radios
SPAN [THO 12]	The MITRE Corporation	Small communities worldwide	WiFi ad-hoc network

Table 10.1. *Community projects for ad-hoc networking*

Freifunk ("free radio" in German) was originally conceived as a method of providing communication to remote locations where infrastructure deployment was deemed not commercially feasible. Hence, its main focus was on fixed wireless communications based on WiFi mesh networking. Over the last 10 years, Freifunk evolved into a stable communication technology running under popular wireless networking protocols (including OLSR [CLA 03] and B.A.T.M.A.N [KLE 12] multi-hop routing protocols) that support Internet access.

Village Telco is another fixed WiFi mesh setup that was built to primarily support voice communication and voice messaging in regions where traditional cellular services were not affordable. From its inception, Village Telco has had to use WiFi technology that operates in unlicensed bands in order to keep costs down. To support basic handsets, the project developed analog telephone adaptors based on the OpenBTS [IED 15] technology for software conversions between the GSM protocol stack and the Internet protocols that are used on top of WiFi. At the same time, these adaptors act as repeaters for coverage extension.

Commotion Wireless is yet another WiFi mesh uptake that was purposefully developed to counteract the actions of regimes that tried to suppress freedom of speech. Similar to Village Telco, it also uses OpenBTS to support both feature handsets and WiFi-enabled devices. It has been successfully used by the Red Hook Initiative (RHI) for WiFi access after Hurricane Sandy. Luckily, the RHI building was one of the very few places that had managed to keep power and thus the network operations were maintained.

Serval is a more recent incarnation of Village Telco in which ad-hoc networking is persued to enable communications between mobile devices. Serval is being developed for use by smartphone devices using the Android operating system.

Ideally, the software package can be downloaded from any other device that already runs Serval, is automatically installed to the device, and becomes immediately operational. However, the software is not compatible with all devices and thus custom mesh routers are being employed to establish communication and extend network coverage, similar to the case of Village Telco.

Finally, Smart Phone Ad-hoc Network (SPAN) is specifically designed for ad-hoc communication between smartphone devices. Like Serval, it only works on particular smartphone devices running Android and it also makes use of Internet protocols to support communication services. The novelty of SPAN is that it implements a generic networking architecture on top of which any routing algorithm can be implemented. At the same time, the SPAN operations are transparent to all layers above and thus no software updates are necessary for any of the applications installed on the devices. Of course, a number of other projects of similar nature also exist, including Project Byzantium (http://project-byzantium.org/), LifeNet (www.thelifenetwork.org/) and PodNet [HEL 10].

Both Freifunk and Village Telco have been created to support fixed access networks and thus reliability is their biggest concern. Commotion Wireless, Serval and SPAN aim for immediate network availability to allow instantaneous communication. Hence, network survivability is not their primary concern. Furthermore, all these projects are similar in the sense that they try to offer a common ground on top of which Internet protocols can run. Unfortunately, it is becoming increasingly clear that such an approach: (1) wastes valuable energy that is crucial for network survivability (especially for battery-operated mobile devices) (2) entails too much signaling overhead and (3) suffers from scalability issues [DAR 13].

10.3. Beaconing-based proximate communication

Recognizing the issues with the implementation of Internet protocols in the mobile domain, standards organizations and equipment vendors are currently in the process of amending popular wireless technologies to support beaconing-based communication. Beaconing enables mobile devices to duty-cycle between the active and sleep states to attain energy efficiency. The actual beacons sent out intermittently ensure device discovery and allow communication.

Table 10.2 lists the respective technologies that are currently in the process of being standardized or that already support beaconing functionality. ZigBee is a well-established standard that incorporates the beacon mode for energy efficient operation. It has been extensively used in wireless sensor networks particularly within the home ecosystem. Bluetooth is another protocol stack that operates in beacon mode to enable low energy operation of devices that follow predefined communication profiles. Both these standards can sustain operational autonomy for weeks to months at the expense of limited communication range.

Technology name	Beaconing Moniker	Applicability	Operation
ZigBee	Beacon mode [ALL 05]	Inter-appliance connectivity	Beacons used to maintain a synchronized network and allow internetworking
Bluetooth	Bluetooth Low Energy (BLE), Bluetooth-Smart [DEC 14]	Fitness trackers, well-being wearables, home entertainment	Profiling-based operation for sending and receiving short messages at predetermined intervals
Wireless Fidelity (WiFi)	WiFi-Direct [WIF 10]	Ad-hoc communication between devices and between users	Mimic WiFi access points in software so as the complete system can operate like traditional WiFi hotspots. Hence, only one device needs to be WiFi-Direct compliant while legacy devices connect to this software access point
	DSRC / ETSI ITS [ETS]	Intervehicle Communication	IEEE 802.11p technology for fast connection establishment between vehicles for predefined message broadcasting
Long-Term Evolution (LTE)	LTE-Direct [3GPP]	Device-to-device communication for public safety service, data traffic offloading, proximity services (e.g. location-based advertising)	Operates under LTE spectrum and may utilize the network timing, allocation mechanisms and user authentication

Table 10.2. *Beaconing technology enablers*

Conversely, WiFi-Direct/DSRC and LTE-Direct achieve communication distances that exceed 100m in range, but at a slightly higher energy expense. WiFi-Direct was designed to enable ad-hoc communication for consumer electronics. DSRC (Dedicated Short-Range Communications) is a highly anticipated vehicular communication technology to improve safety and efficiency on the road. For that protocol, beacons are broadcasted at a rate of 1–10 Hz depending on the wireless channel congestion [GOZ 12]. Nearby vehicles that listen to the wireless broadcast channel, process all received beacons to raise situational awareness, and when needed, trigger preventive actions to avoid collision.

LTE-Direct is the latest standardization method to provide proximate communication using the global LTE cellular connectivity standard. It consists of two basic phases: (1) the discovery of proximate peers and applications and (2) communication between proximate services. In the discovery phase, service "expressions" are used to advertise the presence of peers and their services/applications/context. These expressions are mapped to 128 bit broadcasted beacons at the physical layer (referred to as "expression codes"). LTE-Direct is a synchronous system and thus these expressions are exchanged within pre-scheduled discovery resources on the LTE uplink spectrum (assigned either by the radio access network or in the absence of it distributedly). LTE-Direct supports both operator-controlled and standalone ad-hoc communication for a variety of innovative services, including emergency services. Already, a memorandum of understanding has been agreed between TETRA + Critical Communications Association (TCCA) and the National Public Safety Telecommunications Council for the development of LTE-Direct solutions for emergency communication [CRI 12] [GPP 12].

All these beaconing technologies ensure successful communication between adjacent devices (i.e. Physical and Data Link layer functionality) on top of which any networking solution can be applied. Under this regime, Internet protocols seem to encumber ad-hoc networks with excessive signaling and unnecessary energy waste, defeating the purpose of the beaconing-based approach. Hence, alternative networking solutions are necessary to benefit from the full potential of beaconing-based communications.

In the following sections, a number of functionalities are derived that adhere to decentralized, distributed and direct principles of communication, and are perfectly aligned with the beacon mode of communication to provide resilient telematics in ad-hoc PSNs.

10.4. Beaconing-based networking

As outlined above, beaconing-based networking refers to the networking paradigm where all network-related operations take place locally between connected devices. As an example, devices make data forwarding decisions based solely on information received from adjacent nodes without any additional information. The immediate benefit is that networking information can be broadcasted within beacon messages, thus drastically reducing communication overhead and energy consumption. Moreover, no network-wide operations are necessary, further curbing redundant signaling and processing. Evidently, both these features are particularly appealing to ad-hoc PSNs, where network survivability is a top priority.

Core functionalities that are critically important for ad-hoc PSNs include:

– intelligent alert messaging;

– location identification;

– situation awareness;

– evacuation navigation.

The first two functionalities are concerned with knowledge production and dissemination; both of which are urgently needed in search-and-rescue operations.

The third functionality, situational awareness, refers to the need of all actors to learn what is happening and to assess the urgency of subsequent actions. Finally, evacuation strategies provide meta-data to distress survivors in order to help them escape danger. In the following sections, each of these core functionalities are discussed in the context of beaconing-based networking.

10.4.1. *Intelligent alert messaging*

One of the primary goals of an ad-hoc PSN is to allow stranded survivors to radio for help, even though there is no way to ascertain the best direction to forward alert notifications. Clearly, flooding the network with these notifications will cause excessive energy waste and penalize network survivability. It would thus be more favorable to selectively transmit alerts towards denser regions of the network (i.e. where many devices are present) and towards those regions where devices have plentiful residual battery capacity. With regards to the first requirement, there is a greater chance that first responders will be able to contact devices that lie within densely populated regions of the disaster area, while the second requirement ensures that alert notifications survive within the network for longer. Hence, each device needs to make sense of its significance within the network with respect to the regional density and energy reserves.

To do so, metrics need to be computed for each device to ascertain the significance of each device within its neighborhood. Such metrics can be obtained, for example, from complex network theory including eigenvector centrality, betweenness centrality and many more [NEW 10]. Eigenvector centrality is extensively used in complex network theory to assess the significance of nodes in a network based on the eigenvector of the network adjacency matrix. Local eigenvector centrality (LEC), a version of eigenvector centrality that is based only on the connectivity of a particular device and its direct neighbors, can be effortlessly used in ad-hoc PSNs. The LEC metric assigns higher values to devices with many neighbors or to those devices that have neighbors with high LEC values themselves. It uses a single-value parameter that is recursively updated by an autoregressive function. LEC consists of two parts: the residual energy of the particular device and a normalized sum of the LECs of each adjacent device. The recursive computation of the normalized values converges after only a few cycles, since LEC is only exchanged between neighboring devices. Evidently, this single-value parameter can be easily incorporated within beaconing messages at the expense of only a marginal increase in overhead. Moreover, using LEC, natural regions (clusters) are formed as shown in Figure 10.2(a).

By comparing their own LEC values with those of their neighbors, each device can infer on its topological position. A device with an LEC value higher than that of its neighbors is set to be the cluster head. A device with an LEC value lower than that of all its neighbors can be either a leaf or a bridge node while all other devices are intermediate nodes within a particular cluster (a number of these cluster formations are illustrated in Figure 10.2(a)). By simply forwarding alerts towards nodes with higher LEC values, all alert messages will end up at cluster heads which are by definition nodes in the network that are both highly connected and around neighbors with ample battery reserves. Doing so improves the accessibility to these messages and their survivability over longer time periods.

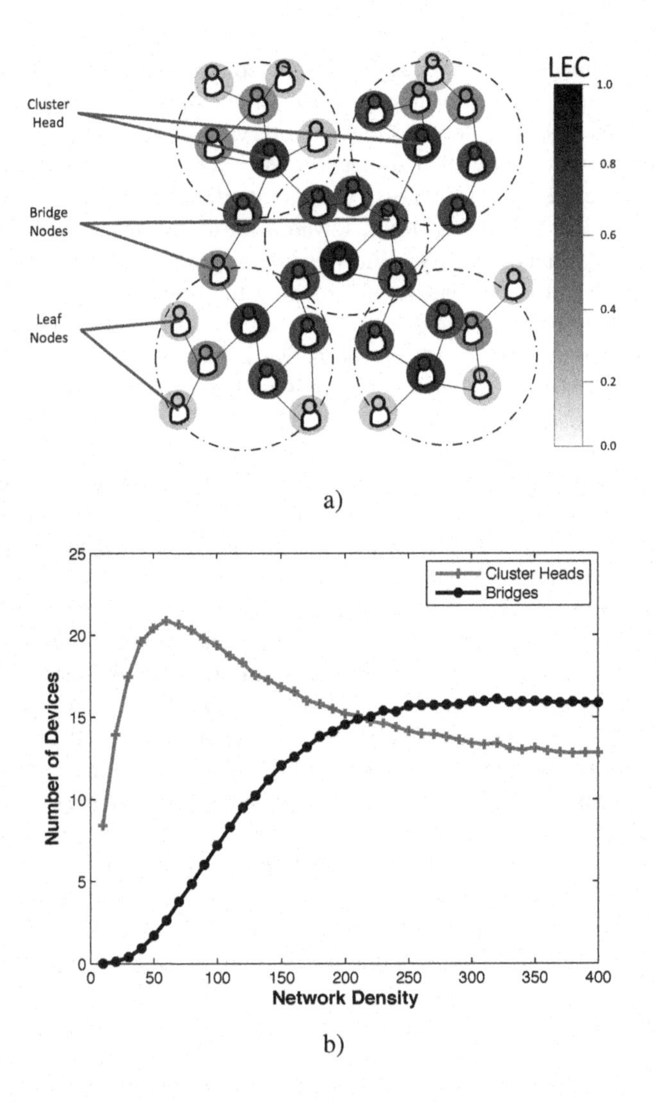

Figure 10.2. *a) Regions formed based on LEC values and b) number of cluster heads and bridges at various network densities*

The benefit of using such topological mapping is that informed forwarding strategies can now be realized as discussed in [KOL 14]. It should be evident from the above description that when each device forwards alerts towards those neighbors with higher LEC values, then at least a single cluster head will receive a particular alert message. If, in addition, each alert notification is sent to devices that act as either cluster heads or bridge nodes, then all alerts will eventually be received by all cluster heads. As a result, a good dispersion of alert notifications will be achieved without flooding the network with excess data.

Figure 10.2(b) illustrates the resulting number of cluster heads and bridge nodes formed within an ad-hoc PSN of varying density of devices. The disaster area is assumed to be a circular region of 1 km radius. All devices are assumed to operate with WiFi-Direct technology and are at random locations within the disaster area. As shown in the figure, for sparse topologies the number of cluster heads increases while for denser topologies the number of clusters converges and the total number of cluster heads stabilizes. At the same time, the number of bridge nodes increases monotonically with the total number of clusters formed within the PSN and it reaches a plateau when the total number of clusters stabilizes. Noticeably, the total number of cluster heads and bridge nodes are only a small fraction of the total number of devices forming an PSN, and thus forwarding alerts to these particular devices has a significant energy efficiency potential. At the same time, alerting these devices of all distress messages improves reachability with the potential to expedite search-and-rescue.

10.4.2. *Device localization*

Location information is undoubtedly one of the most critical pieces of data that are offered to survivors and emergency responders. An immediate benefit of localization is that each device can automatically add geolocation tags to all alert notifications produced. Doing so will aid rescue operations and help with victim triage [SAK 13]. At the same time, location information can advance forwarding strategies for better dispersion of alert messages along the disaster area.

The majority of handsets coming out in the market have built-in positioning capabilities that are primarily based on GPS receivers. GPS signals, however, are only strong enough to be received outdoors and fail to provide a positioning fix indoors. At the same time, there are many indoor localization solutions that achieve positioning through the site's fixed communication network (e.g. WiFi deployment) which is prone however to damage and destruction especially during disasters.

Alternatively, each device can estimate its position from location information broadcasted by neighboring nodes. Initially, a number of stranded survivors may be able to obtain a GPS fix. In turn, adjacent devices can get a crude estimate of their location and recursively correct that estimate as more location information is diffused in the network. It is evident that the beaconing mechanisms within emergency response networks can effortlessly support this recursive operation.

The SNAP (Subtract on Negative Add on Positive) localization algorithm is one recursive solution that can be implemented in a decentralized, distributed and direct manner [MIC 09]. With SNAP, each device is required to broadcast its own location estimate and location estimates of its neighbors at intermittent intervals. Using SNAP, an unlocalized device estimates its position based on location estimates received from its neighbors. In successions, this estimate is updated as new information becomes available. With regards to the communication overhead of SNAP, each location fix is completely characterized by 63 bits and thus the total beaconing overhead is equal to $(N + 1) \times 63$ bits, where N is the total number of neighbors to a particular device.

To compute a location estimate, SNAP assumes that each device lies within a quantized field. Initially, every block of this field is set to zero. For every position fix contained within the broadcasted beacons, SNAP draws a coverage area (e.g. rectangular) twice the maximum communication range (60 m for BLE, 200 m for WiFi-Direct and 500 m for LTE-Direct). These coverage areas are centered around each of the collected position estimates. For all blocks that fall within the area of the direct neighbors of a particular device, a 1 is added to the block entries. For all other coverage areas (associated with devices that are not direct neighbors but are instead two-hop neighbors), a 1 is subtracted from the block entries. Finally, the block entries are summed up and a location estimate is computed as the centroid of all blocks that attain the maximum block values.

The most important components of SNAP are illustrated in Figure 10.3. A positive unit value is considered for all grid blocks within the area covered by the estimated communication range while all two-hop neighbors impose a negative unit value across grid blocks falling with their coverage, as shown in Figure 10.3(a). The accumulated contributions (both positive and negative values) of the grid blocks that overlap on the field are presented in Figure 10.3(b). For the particular device that runs SNAP, a location estimate is set to be the centroid of the blocks with the highest block values (shown as shaded in the figure). Importantly, the work in [MIC 09] demonstrates that SNAP is in essence a recursive maximum-likelihood estimation algorithm that obtains adequate localization accuracy with low computational cost.

Figure 10.4 provides indicative localization levels that are expected of SNAP for varying network densities. As before, the underlying disaster area spans across a circular region of radius 1 km and WiFi-Direct is assumed to be used by all devices. For SNAP, a rectangular coverage area is assumed with grid blocks of 50 m^2. In Figure 10.4(a), any device is assumed localized if the localization error falls below 100m. Such a localization error would be unusable in the general case, but for emergency response this information does have merit. As shown in the figure, a number of unlocalized nodes can obtain a location estimate even in sparse topologies. Figure 10.4(b) further shows the localization error of SNAP for an ad-hoc PSN that consists of 400 devices while varying percentages of devices are able to obtain a GPS fix. As shown in this figure, SNAP uses the information diffused in the network to correct location estimates, bringing the error down to a few tens of meters after only a small number of iterations.

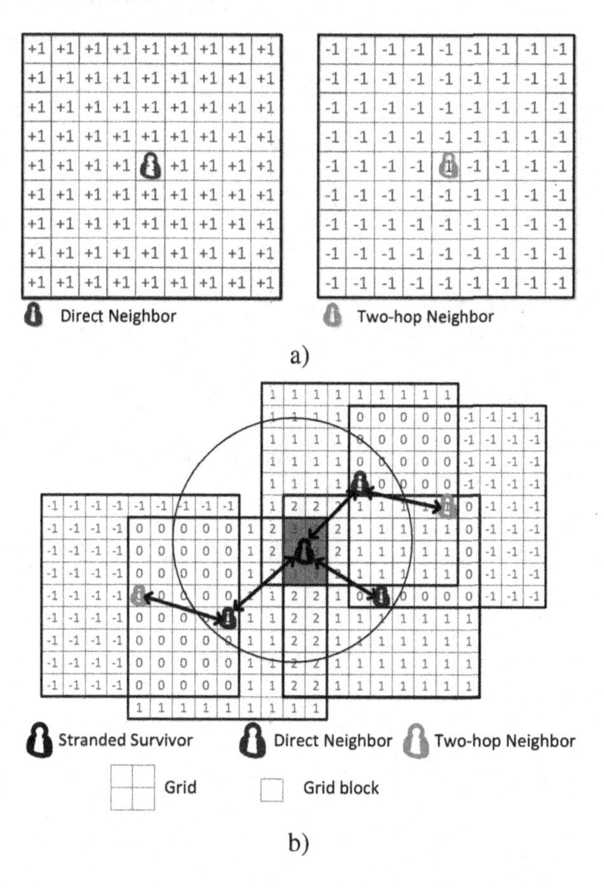

Figure 10.3. *SNAP node localization*

10.4.3. *Situation awareness*

As a result of the aforementioned two functionalities, a network of localizable components can be formed. Components refer to patients, first responders, bystanders, any kind of equipment (e.g. vehicles) and resources (e.g. clothing) that could prove valuable during the response to an emergency. In this way, alert messages can be geo-tagged and timestamped before they are forwarded in the network so as to improve situational awareness. In addition, a plethora of other data sets can be made available to assist response efforts in the field.

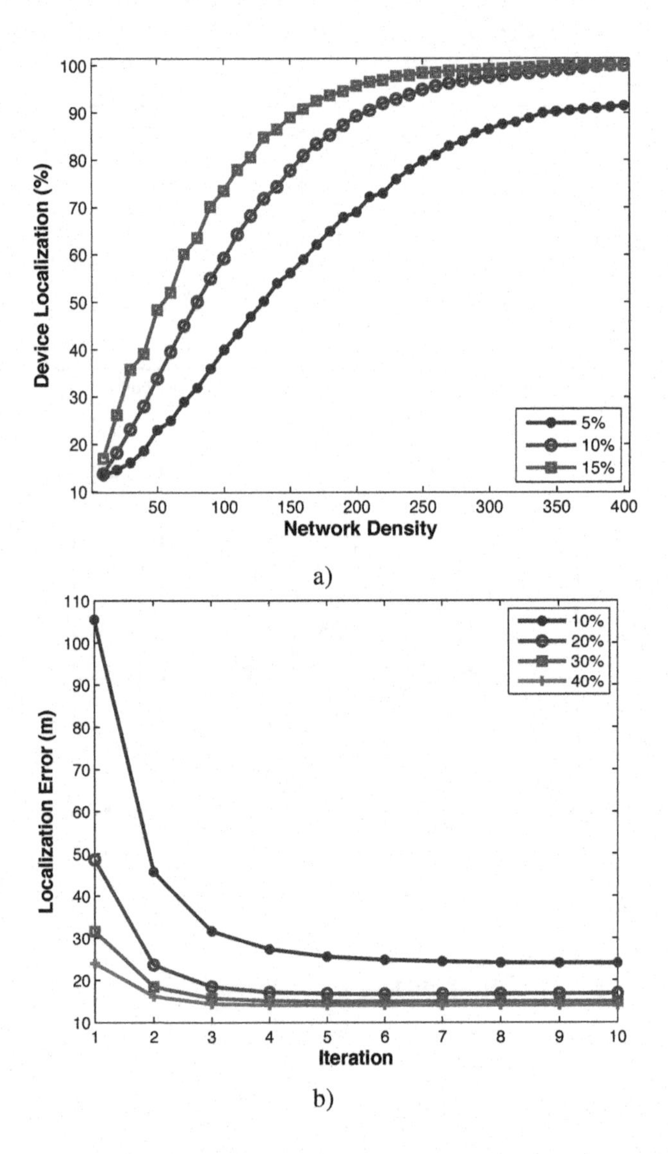

Figure 10.4. *a) Percentage of devices localized for varying network densities and b) localization error for a PSN of 400 devices for varying percentage of devices that have obtained a GPS fix*

Patient's physiological parameters (including heart rate, blood pressure, surrounding temperature, etc. [CUR 08] [LEN 11]) and triage status (including level and color [KOJ 11] [MAR 11]) can be made accessible through electronic forms on mobile devices and wearables [TIA 08]. At the same time, profiles of all other actors can be used to inform of the available competencies and skills in the vicinity. These types of information can be exchanged between actors in the network for classification and matching. Similarly, equipment and resource forms can be used to disseminate information on the type, availability and quantity of potentially helpful equipment and resources in the field. Obviously, the increase in data sharing requirements (associated with the exchange of the additional information), creates further challenges in the underlying network.

A promising approach for sharing these common data pools in ad-hoc PSNs is through the use of principles developed by peer-to-peer (P2P) applications [CAT 08]. In addition to efficient data dissemination (achieved by approaches similar to the one discussed in section 10.4.1), a P2P overlay enables efficient content search, sharing of resources and data permanency [LI 13]. These characteristics introduce scalability and fault tolerance in applications and services, both of which are particularly important features of a communications network especially during emergencies [MIL 09]. Moreover, numerous adaptations of basic P2P overlays have already been proposed to address the ad-hoc nature of communication, including location-aware distributed hash tables, and neighborhood-based replica placement and peer selection [BRA 08]. By adapting these solutions to handle the intermittency of beaconing-based approaches, both the P2P characteristics and the resilience of ad-hoc networks that are created instantaneously can be achieved.

10.4.4. *Evacuation navigation*

A shared and up-to-date understanding of the situation enables correct decisions to be taken during an emergency, with evacuation navigation being a top priority. Existing solutions focus on devising *a priori* plans and procedures that are hardcoded on infrastructures. However, as elaborated in the introduction, such infrastructures are susceptible to damage, while unanticipated emergency situations can make pre-planned evacuation routes completely invalid.

In contrast, an ad-hoc PSN can actively support emergency evacuation via distributed decision-making based on actual field information and user cooperation. The work in [PIS 15], for example, investigates such an approach in indoor environments. To compute safe navigation routes, information about the physical distance between people, the distribution of people in an area and the degree of mobility of these people is needed, in addition to detecting incidents and hazards. As emphasized above, such metrics can be computed in a distributed manner using a beaconing message exchanged between neighboring nodes. With this information at hand, a plethora of different algorithms can be used for wayfinding through which safe evacuation can be achieved while mitigating congestion.

Table 10.3 lists the most prominent algorithms that are currently being considered for emergency evacuation. A review of these algorithms (as described in the table) shows that they either depend on input from existing wireless sensor networks or other infrastructure support, including computing clouds for heavy computations. More important, however, is the fact that preliminary work on purely ad-hoc solutions has demonstrated that the complexity of achieving this functionality is very high. As indicated in [GOR 12], efficient and effective evacuation solutions in ad-hoc PSNs is hindered by problems with incomplete and incorrect data dissemination, and communication delays that cause great problems while deciding on safe evacuation strategies. Hence, achieving this functionality in practical scenarios is still a great challenge.

Algorithm	Description
Li *et al.* [LI 03]	A distributed algorithm based on artificial potential fields that computes minimum exposure paths
Tseng *et al.* [TSE 06]	An algorithm based on TORA that alters weights to navigate away from hazards
SOS [ZHA 11]	A distributed flow-based algorithm that computes safe paths while taking into account human dynamics
Gelenbe *et al.* [GEL 14]	A cloud-based algorithm that combines social potential fields and a cognitive packet network
CANs [WAN 15]	Uses a level set method to track the evolution of the exit and the boundary of the hazardous area. Local status updates are used for dynamic emergency updates

Table 10.3. *Existing evacuation navigation algorithms*

10.5. Concluding remarks

Beaconing-based networking is becoming an integral part of popular wireless technologies. As shown in this chapter, this mode of operation offers the opportunity to realize novel networking solutions that are decentralized, distributed and direct in nature. Efficiency, robustness and resilience are some of the virtues of this networking solution that has the potential to make ad-hoc PSNs truly practical. This chapter approaches key functionalities that can be supported by mobile devices spontaneously forming ad-hoc PSNs, giving hard evidence of the feasibility and applicability of this complementary networking solution in emergency scenarios.

10.6. Bibliography

[3GPP] 3GPP, "Delivering public safety communications with LTE", available at: http://www.3gpp.org/news-events/3gpp-news/1455-Public-Safety.

[ADE 11] ADEYEYE M., STEPHEN P.G., "The Village Telco project: a reliable and practical wireless mesh telephony infrastructure", *EURASIP Journal on Wireless Communications and Networking*, vol. 1, no. 78, available at: http://www.shuttleworthfoundation.org/, 2011.

[ALL 05] ALLIANCE Z., "ZigBee Specification", ZigBee Standards Organisation, available at: http://www.zigbee.org, 2005.

[BRA 08] BRADLER D., KANGASHARJU J., MUHLHAUSER M., "Evaluation of peer-to-peer overlays for first response", *IEEE International Conference on Pervasive Computing and Communications*, March 2008.

[CAT 08] CATARCI T. *et al.*, "Pervasive software environments for supporting disaster responses", *IEEE Internet Computing*, vol. 12, no. 1, pp. 26–37, 2008.

[CHE 10] CHEN G., HU A., SATO T., "A scheme for disaster recovery in wireless networks with dynamic ad-hoc routing", *ITU-T Kaleidoscope: Beyond the Internet? – Innovations for Future Networks and Services*, available at: http://ieeexplore.ieee.org/stamp/stamp.jsp?tp=&arnumber=5682120&isnumber=5682114, Pune, pp. 1–6, 2010.

[CLA 03] CLAUSEN T., JACQUET P., "Optimized link state routing protocol", available at: http://www.olsr.org/, IETF RFC 3626, 2003.

[COR 10] CORLEY A.M., "Why Haiti's cellphone networks failed", *IEEE Spectrum*, http://spectrum.ieee.org/telecom/wireless/why-haitis-cellphone-networks-failed, 2010.

[CRI 12] CRITICAL COMMUNICATIONS BROADBAND GROUP, "Additional information: group communications & proximity-based services", *Liaison Statement*, available at: http://3gpp.org/ftp/tsg_sa/TSG_SA/TSGS_57/Docs/SP-120456.zip, 2012.

[CUR 08] CURTIS D.W. *et al.*, "SMART-an integrated wireless system for monitoring unattended patients", *Journal of the American Medical Informatics Association*, vol. 15, no. 1, pp. 44–53, 2008.

[DAR 13] DARPA, "Novel Methods for information sharing in large-scale mobile ad-hoc networks", *DARPA-SN-13-35 Request for Information*, 2013.

[DEC 14] DECUIR J., "Introducing bluetooth smart: part 1: a look at both classic and new technologies", *IEEE Consumer Electronics Magazine*, vol. 3, no. 1, available at: http://www.bluetooth.com/Pages/Bluetooth-Smart.aspx, 2014.

[ETS] ETSI, "ETSI, Intelligent Transport Systems (ITS) Standards", available at: http://www.etsi. org/index.php/technologies-clusters/technologies/intelligent-transport/dsrc.

[FRE 16] FREIFUNK, "Förderverein Freie Netzwerke e.V.", available at: http://freifunk.net/, 2016.

[GAR 13] GARDNER-STEPHEN P. *et al.*, "The serval mesh: a platform for resilient communications in disaster & crisis", *IEEE Global Humanitarian Technology Conference*, 2013.

[GEL 14] GELENBE E., BI H., "Emergency navigation without an infrastructure", *Sensors*, vol. 14, no. 8, pp. 474–477, 2014.

[GEO 10] GEORGE S.M. *et al.*, "DistressNet: a wireless ad hoc and sensor network architecture for situation management in disaster response", *IEEE Communications Magazine*, vol. 48, no. 3, pp. 128–136, 2010.

[GER 14] GERETY R., MOSES B.N., GUNN A. *et al.*, "Commotion wireless as a community technology", *New America Foundation*, available at: http://oti.newamerica.net/, 2014.

[GOR 12] GORBIL G., GELENBE E., "Resilient emergency evacuation using opportunistic communications", *Computer and Information Sciences III*, pp. 249–257, 2012.

[GOZ 12] GOZALVEZ J., SEPULCRE M., BAUZA R., "IEEE 802.11p vehicle to infrastructure communications in urban environments", *IEEE Communications Magazine*, vol. 50, no. 5, pp. 176–183, 2012.

[GPP 12] 3GPP, "WID for group communication system enablers for LTE", TSG SA1 TD SP-120421 (GCSE_LTE), Report, 2012.

[HEL 10] HELGASON O.R., "A mobile peer-to-peer system for opportunistic content-centric networking", *ACM SIGCOMM MobiHeld Workshop*, available at: http://www.podnet.ee.ethz.ch/, 2010.

[IED 15] IEDEMA M., "Getting started with OpenBTS", *OReilly Media*, available at: http://openbts.org/, 2015.

[ITU 13] ITU-D, "The world in 2013: ICT facts and figures", *ITU Telecommunication Development Bureau*, available at: http://www.itu.int/en/ITU-D/Statistics/Pages/facts/default.aspx, 2013.

[FCC 06] FCC, "Katrina panel, independent panel reviewing the impact of hurricane Katrina on communications networks", *Report and Recommendations of the Independent Panel Reviewing the Impact of Hurricane Katrina on Communications Networks*, available at: https://www.fcc.gov/pshs/docs/advisory/hkip/karrp.pdf, 2006.

[KLE 12] KLEIN A., BRAUN L., OEHLMANN F., "Performance study of the better approach to mobile ad hoc networking (B.A.T.M.A.N.) protocol in the context of asymmetric links", *IEEE International Symposium on a World of Wireless, Mobile and Multimedia Networks*, available at: www.open-mesh.org/, 2012.

[KOJ 11] KOJIMA H., NAGAHASHI K., OKADA K., "Proposal of the disaster-relief training system using the electronic triage tag", *IEEE International Conference on Advanced Information Networking and Applications*, March 2011.

[KOL 14] KOLIOS P. *et al.*, "Explore and exploit in wireless ad hoc emergency response networks", *IEEE International Conference on Communications*, 2014.

[KWA 12] KWASINSKI A., Hurricane sandy effects on communication systems, Preliminary Report PR-AK-0112-2012, 2012.

[LEN 11] LENERT L.A. *et al.*, "Design and evaluation of a wireless electronic health records system for field care in mass casualty settings", *Journal of the American Medical Informatics Association*, vol. 18, no. 6, pp. 842–852, 2011.

[LI 03] LI Q., ROSA M.D., RUS D., "Distributed algorithms for guiding navigation across a sensor network", *International Conference on Mobile Computing and Networking*, 2003.

[LI 13] LI L., YANFANG J., YUE Z., "A survey on P2P file sharing algorithms over MANETs", *Consumer Electronics Times*, vol. 2, no. 2, pp. 109–115, 2013.

[LIU 13] LIU S., STRIEGEL A.D., "Exploring the potential in practice for opportunistic networks amongst smart mobile devices", *International Conference on Mobile Computing and Networking*, 2013.

[MAR 11] MARTÍN-CAMPILLO A., MARTÍ R., YONEKI E., "Electronic triage tag and opportunistic networks in disasters", *ACM Special Workshop on Internet and Disasters*, 2011.

[MAS 11] MASE K., "How to deliver your message from/to a disaster area", *IEEE Communications Magazine*, vol. 49, no. 1, pp. 52–57, 2011.

[MIC 09] MICHAELIDES M., LAOUDIAS C., PANAYIOTOU C., "Fault tolerant detection and tracking of multiple sources in WSNs using binary data", *IEEE Conference on Decision and Control*, 2009.

[MIL 09] MILLAR G.P., RAMREKHA T.A., POLITIS C., "A peer-to-peer overlay approach for emergency mobile ad hoc network based multimedia communications", *International ICST Mobile Multimedia Communications Conference*, 2009.

[NEW 10] NEWMAN M.E.J., *Networks: An Introduction*, Oxford University Press, 2010.

[NTT 12] NTT DOCOMO, "Measures for recovery from the great east Japan earthquake using NTT Docomo R&D technology", *NTT Docomo Technical Journal Editorial Office*, vol. 13, no. 4, pp. 96–106, 2012.

[PIS 15] PISCITELLO A., "Danger-system: exploring new ways to manage occupants safety in smart building", *IEEE World Forum on Internet of Things*, 2015.

[SAK 13] SAKURAI M. *et al.*, "Sustaining life during the early stages of disaster relief with a frugal information system: learning from the great east Japan earthquake", *IEEE Communications Magazine*, vol 52, no. 1, pp. 176–185, 2013.

[SUZ 12] SUZUKI N. *et al.*, "Using SOS message propagation to estimate the location of immobilized persons", *International Conference on Mobile Computing and Networking*, 2012.

[THO 12] THOMAS J. *et al.*, "Off grid communications with android meshing the mobile world", *IEEE Conference on Technologies for Homeland Security*, 2012.

[TIA 08] TIA G. *et al.*, "The advanced health and disaster aid network: a light-weight wireless medical system for triage", *IEEE Transactions on Biomedical Circuits and Systems*, vol. 1, no. 3, pp. 203–216, 2008.

[TSE 06] TSENG Y., PAN M., TSAI Y., "A distributed emergency navigation algorithm for wireless sensor networks", *IEEE Computers*, vol. 39, pp. 755–62, 2006.

[WAN 15] WANG C., LIN H., JIANG H., "CANS: towards congestion-adaptive and small stretch emergency navigation with wireless sensor networks", *IEEE Transactions on Mobile Computing*, doi:10.1109/TMC.2015.2451639, 2015.

[WIF 10] WIFI ALLIANCE, "Wi-Fi Peer-to-Peer (P2P) Technical Specification v1.1", Wifi Alliance P2P Task Group, available at: http://www.wi-fi.org/discover-wi-fi/wi-fi-direct, 2010.

[ZHA 11] ZHAN A., WU F., CHEN G., "SOS: a safe, ordered, and speedy emergency navigation algorithm in wireless sensor networks", *International Conference on Computer Communications and Networks*, 2011.

11

ICN/DTN for Public Safety in Mobile Networks

11.1. Introduction

Long-Term Evolution (LTE) is a new communication standard developed by the Third Generation Partnership Project (3GPP). Currently, LTE is becoming a 4G reference architecture due to its widespread adoption among leading operators of mobile telecommunications. LTE is therefore foreseen as an important foundation for future 5G networks. In the shift towards 5G, several open issues have to be worked out in LTE. They emerge due to severe requirements put on the infrastructure of the future networks. First, mobile users will expect high capacity channels, in which capacity is measured in several Gbps. Second, new applications will be considered

---

Chapter written by Eryk SCHILLER, Eirini KALOGEITON, Torsten BRAUN, André GOMES, and Navid NIKAEIN.

with high densities of connected devices. Third, 5G networks will have to accommodate new types of connected devices such as household appliances, meters and connected cars. Fourth, direct device-to-device (D2D) communication will have to be formulated for sharing information in a local context. In the broader view, network-based communication in the licensed band can provide enhanced Quality of Service (QoS) for D2D scenarios. Fifth, extreme reliability (e.g. medical applications) and ultra-low latencies (e.g. VANET applications) have to be considered. Sixth, for Internet of Things (IoT) applications, communication with high energy efficiency is required. Finally, in current mobile networks, we are approaching Shannon's capacity limit. Therefore, an enhanced channel capacity shall be provided through the adoption of a new spectrum range. According to the aforementioned picture, 5G will be a holistic ecosystem providing connectivity in a wide range of application use cases. It is therefore natural to seamlessly integrate Public Safety (PS) applications with 5G using LTE as a starting point.

LTE has been selected by the National Public Safety Telecommunications Council (NPSTC) in the USA as a basis for PS. Other regions of the world, such as the European Union, will most probably adapt LTE for PS as well. Currently existing PS systems such as Project 25 (P25) and Terrestrial Trunked Radio (TETRA) are already reliable in providing voice communication; however, new high bandwidth applications in PS can only be provided by LTE. Currently, LTE does not natively support PS, as it was designed to support commercial cellular networking. It does not allow for a required level of reliability, security and confidentiality. Moreover, device-to-device communication is also not appropriately taken into account. Hence, a new research area is emerging on an appropriate adaptation of LTE towards PS networking. 3GPP already addressed several of these issues in their studies such as device-to-device communication, Evolved Universal Terrestrial Radio Access Network (E-UTRAN) and Mission-Critical Push-to-Talk (MCPTT). 3GPP worked out specific properties of both the User Equipment (UE) and the evolved NodeB (the LTE base station eNB) that have to be taken into account in the provision of PS applications with respect to various availability levels of E-UTRAN. More specifically, 3GPP consider an isolated E-UTRAN scenario, in which an eNB operates with no access or limited connectivity to the LTE core (EPC). In such situations, there is a need for rapid provisioning of the LTE network. The 3GPP does not study, however, how distributed disconnected (isolated) eNBs exchange information. This is left for vendors to implement their own proprietary solutions [FAV 16].

Delay Tolerant Networking (DTN) and Information Centric Networks (ICN) can provide added value to LTE. Over the years, DTN has become an emerging push-based paradigm for challenge networks. When a disaster occurs, communication has to be reestablished to ensure PS in areas in which infrastructure is limited, power supplies might be damaged and/or the network is disconnected from its main core. DTN is an Internet architecture that overcomes technical difficulties that may exist in these challenge environments. In a disaster scenario, we must assume that nodes may be disconnected from their network and/or from each other. Hence, it is possible that information may not be immediately delivered to the destination due to a momentary lack of end-to-end connectivity.

The main advantage of DTN is that it enables communication in intermittent networks through its store-carry-and-forward mechanism. Every node in the network could store a message in its buffer when no connection is available. The node stores the message and can move to any direction, until a connection reappears. Then, the node forwards the message to other nodes so that the message can gradually approach the destination. In addition, another advantage of DTN includes the potential of working in heterogeneous networks (IP/non-IP) using the bundle protocol. Bundle protocol enables messages to be of variable sizes and enables multi-hop communication in order for a message to reach its final destination. This opportunistic model could change its routing decisions depending on the network topology. These two main advantages of DTN are exploited in PS to ensure that a message reaches rescue teams in disaster scenarios.

In addition, another concept addressing PS in disaster situations is ICN based. ICN provides a pull strategy for content retrieval using content description for addressing purposes. It is radically different from current networks, which use endpoint identifiers to locate the content. The store-carry-and-forward mechanism is an intrinsic property of ICN, since it does not rely on end-to-end communication, but instead communication is established based on the content of the exchanged messages and not on the location of the host. It can therefore support intermittent connectivity in a catastrophic environment, i.e. scenarios in which isolated nodes have to communicate with rescue teams.

In conclusion, the aforementioned advantages of 5G and DTN/ICN make their combination a perfect candidate to ensure PS in catastrophic scenarios. The rest of this chapter is structured as follows: in section 2, we present related work about MEC systems and DTN/ICN architectures; section 3 presents the proposed MEC system architecture; in section 4, we illustrate an example implementation of the MEC architecture, and finally we conclude in section 5.

11.2. Related work

In Figure 11.1, we depict a simplified schematic of a 4G Mobile Network Operator.

It contains evolved base stations (eNB), Mobility Management Entity (MME), Home Subscriber Server (HSS), Serving Gateway (SGW) and Packet Data Network Gateway (PGW). An eNB is a base station that provides a Radio Access Network (RAN) towards end users operating User Equipment (UE). The RAN is based upon

Evolved Universal Terrestrial Radio Access (E-UTRA) using Orthogonal Frequency-Division Multiple Access (OFDMA) on the downlink and Single-carrier FDMA (SC-FDMA) on the uplink. An eNB communicates with the MME using the S1-AP protocol on the S1-MME interface and with the SGW using GTP-U protocol on the S1-U interface. An MME is a critical network function, which deals with the EPC control plane. Its role is to manage sessions, authentication, paging, mobility, bearers and roaming. It manages an eNB and an S-GW through the S1-MME and S11 interfaces respectively, and communicates with HSS through the S6 interface. An SGW is directly controlled by an MME. It is responsible for routing user plane packets between an eNB (S1-U) and a PGW (S5/S8 interface). It handles user handovers between neighboring eNBs. A PGW is a user plane component, which forwards packets between the LTE network and other packet networks. It is controlled through the S5/S8 interface by the SGW. Moreover, PGW is a mobility anchor for end users. It performs charging and deep packet inspection (lawful interception) and manages Quality of Service (QoS). HSS is responsible for maintaining the user service subscription information.

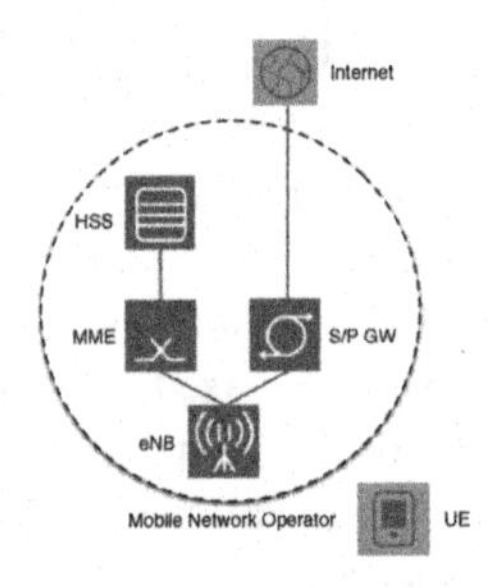

Figure 11.1. *A simplified architecture of the LTE system*

A typical E-UTRAN network can fail in various critical situations which can cause damages to various components of the 5G ecosystem. It results in a destroyed eNB, failed S1 interface disrupting the control and data plane, or damaged EPC. However, in PS communications, the network should guarantee a highly flexible and resilient operation in various situations. Moreover, the PS network should adapt under various circumstances and mobility scenarios to provide a volatile infrastructure of high capacity.

To support local information processing and storage as well as improving robustness and providing low-delay communication to rescue and security teams, the concept of Mobile Edge Computing (MEC) seems a good candidate for provisioning critical systems at the network edge. The MEC concept is normally used to provide local computing and storage solutions at the eNB site. MEC is an extension of Mobile Networks of the future, i.e. 4G and 5G. Due to MEC, content, services and applications can greatly benefit increased responsiveness of the network edge. MEC is also foreseen in 5G as an important technological enabler towards new genres of applications that intelligently combine geolocation, network conditions and radio information all together, to provide enriched services to end users. MEC is currently considered an important enabler of intelligent networking of the future and will be widely spread in the 5G ecosystem.

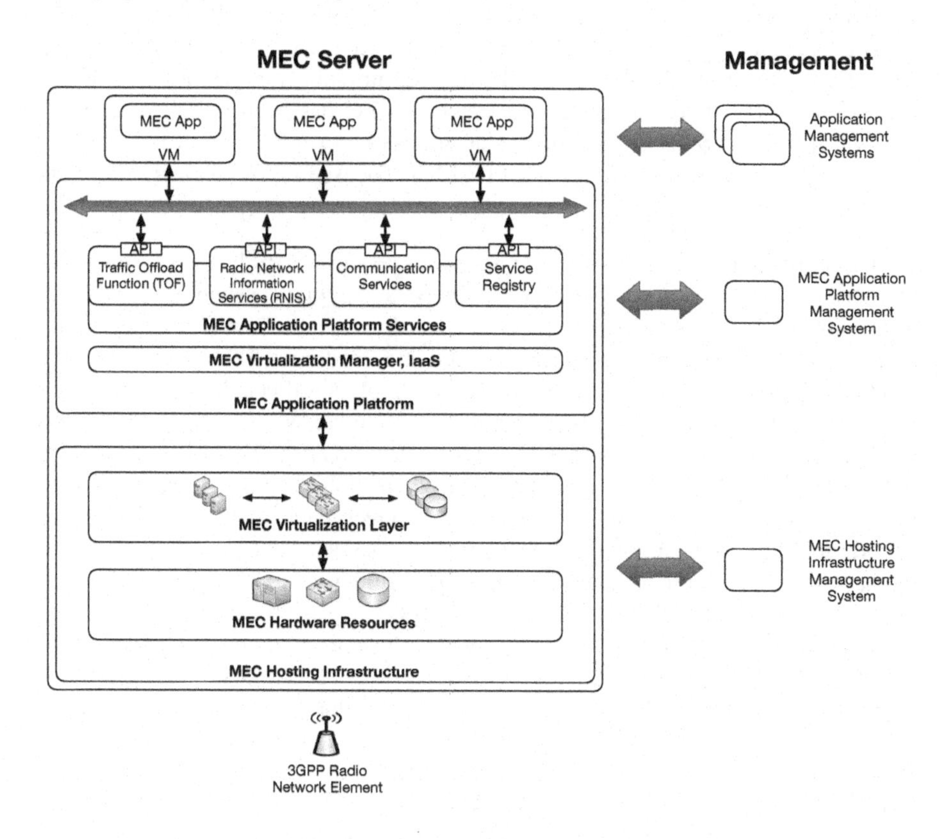

Figure 11.2. *The architecture of the MEC system*

The architecture of a MEC system, which is in the ETSI white paper [PAT 14], is shown in Figure 11.2. At the very bottom of the macro eNB site (see Figure 11.2), we illustrate a 3GPP-compliant radio interface. The MEC hosting infrastructure is provided at the bottom layer of the MEC server. It contains hardware resources (CPU, storage and networking) as well as the MEC virtualization layer. Both hardware and virtualization techniques have to comply with the run-time requirements of MEC applications running on top. The upper MEC application platform consists of the MEC Virtualization Manager and MEC Application Platform Services. The virtualization manager is responsible for the life cycle of MEC

Applications and communicates with the MEC Virtualization Layer. The MEC application platform exposes an API to MEC Applications, i.e. Traffic Offload Function (TOF), Radio Network Information Services (RNIS), Communication Services and Service Registry. First, TOF redirects user traffic towards appropriate MEC applications based on various policies providing application chaining if requested. Second, the MEC application can be provided with radio channel information through RNIS. Third, the Service Registry holds information on currently maintained services. Finally, the Communication Services allow the MEC application to communicate all together. MEC has been investigated in the FP7-MCN project [SOU 16]. In particular, it has been studied how MEC and ICN can complement each other; however, FP7-MCN focused on mobile networks with an operational core.

The utilization of DTN in disaster scenarios to ensure PS has been studied in literature. In [TRO 16], the authors propose a mapping system using a DTN environment in a disaster scenario. Specifically, they use civilians that act as sensor nodes through their mobile devices and collect data. Nodes use DTN to transfer data, reaching computing nodes that perform mapping of the affected area. Moreover, in [UCH 13] the combination of DTN with the Cognitive Wireless Network (CWN) for disaster networks is proposed. Furthermore, Fajardo *et al.* [FAJ 14] implemented a data collection method that uses people and their mobile phones as sensor nodes. People move inside an area of interest to collect data, which is transmitted opportunistically. In addition, the authors propose message aggregation to reduce the message size and minimize the delay. Many DTN studies on a disaster scenario focus on the message transfer, i.e. the routing protocol that is used for message forwarding independently of the network situation (available resources, number of nodes). The authors in [VAH 00] propose the use of epidemic routing to spread messages in the network. In particular, they develop a broadcast technique to exchange messages from a node to all its connections. In addition, in [LIN 03], the Probabilistic Routing Protocol using History of Encounters and Transitivity (PRoPHET) is proposed, where the authors use a learning phase to collect connectivity probabilities in the network. Specifically, each node maintains a possibility of meeting other nodes, and messages are exchanged based on that. Moreover, in [BUR 06], an extension of PRoPHET called "MaxProp" is proposed. It provides acknowledgments. The protocol prioritizes messages with a smaller number of intermediate hops. The protocol also keeps track of the previous message exchange to avoid duplicate transmissions. Spyropoulos *et al.* [SPY 05] introduced a new routing protocol Spray and Wait. This protocol operates two phases. First is spray, in which a certain number of messages are sent to other nodes. After that the wait phase is executed, in which nodes wait to observe whether the previously sent messages reached their destinations. Finally, in [DEM 07], a traditional Link State routing algorithm approach called "Delay Tolerant Link State Routing (DTLSR)" is suggested, where link state announcements are distributed through the network. For each message, the best path to the destination is selected, based on the link queue, the latency and the bandwidth of the link.

Tyson *et al.* [TYS 14] study the utilization of ICN in disaster scenarios. The authors argue that ICN could improve connectivity resilience. This is due to the fact that in an ICN architecture, nodes can explore multiple interfaces at the same time. In addition, ICN does not have to maintain small connection timeouts as in classical networks. Finally, ICN requires no particular underlying network layer, as it creates its own ad-hoc network. Moreover, deploying ICN in a network could improve QoS, as different requests could be treated differently. Furthermore, ICN supports the store-carry-and-forward mechanism, as each node could be equipped with a cache, which is important in disaster scenarios, where connectivity may momentarily disappear. We therefore argue that the integration of DTN/ICN with LTE is an important avenue of research.

11.3. System architecture

In this subsection, we work out a mobile network architecture that provides ICN/DTN network services in the case where a still functional eNB can provide a RAN towards end users (UEs) but it does not have a valid connection to the network core. Note that when a fully functional eNB loses its connection to the core, it stops providing RAN. This situation can be a result of a failed S1 or EPC scenario (see Figures 11.3(a) and 11.3(b)).

Strong protection mechanisms (such as independent power supplies, uninterruptible power supplies, etc.) can be used for MEC infrastructures similarly to as previously used by legacy mobile telephony systems in the case of an eNB. Therefore, a MEC capable macro eNB site could independently operate for a long time after a critical event occurs if the network core fails or the S1 interface malfunctions. The macro eNB site with MEC could therefore successfully provide communicating support to drones and rescue teams in the immediate vicinity even though the network core would not be reachable. In this case, services could be provided in the failure case, according to local communication among attached nodes (in contrast to global communication when the remote core is available). We are therefore targeting the provisioning of disconnected E-UTRAN systems.

A macro MEC-enabled eNB is the main architectural element of the system.

In ordinary situations, when the network core is reachable, our MEC eNB site runs a software based Base-Band-Unit (BBU), which is a software part of an eNB that provides E-UTRAN and communicates with the operator network core to provide mobile access. The primary purpose of the MEC server is to run MEC applications that improve user quality of experience such as caching, online gaming, augmented reality, etc. Due to MEC, the base station can already actively cooperate in the DTN/ICN information dissemination by instantiating DTN/ICN-based services as Virtual Network Functions (VNFs). The primary purpose of this work is to provide DTN/ICN in a disaster situation, when a bundle of a micro LTE core is provided, to run RAN integrated with DTN/ICN.

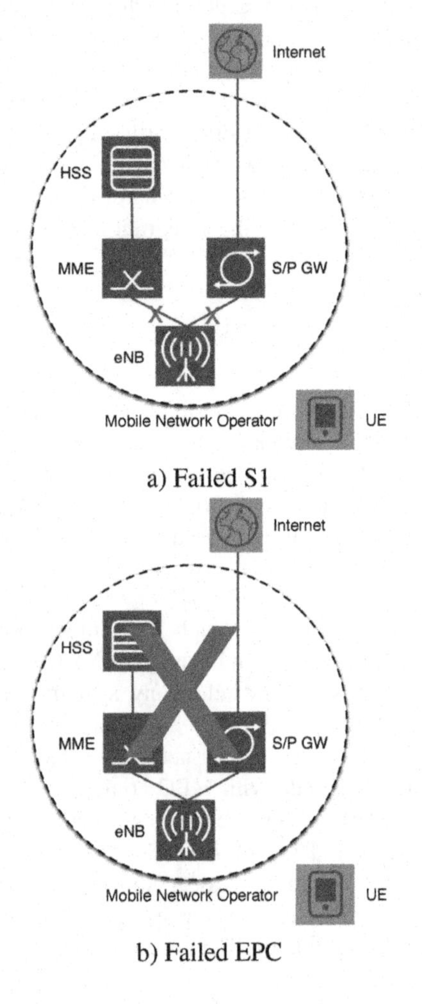

Figure 11.3. *An operational eNB in the disconnected core scenario*

Whenever a local Application Management Unit discovers that an eNB becomes disconnected from the core network, and that the eNB is not able to operate as an isolated eNB, it starts the recovery procedure to provide a new communication service. Such a service is deployed as a bundle of VNFs that define the required network services, namely eNB, local S+P GW, MME and HSS, and PS, and leverage the MEC platform when available to provide a fast network recovery while at the same time offering additional services (see Figures 11.4 and 11.5). User information is maintained either by replicating the HSS database if possible or by provisioning the known IMSI (range) without necessarily the operators key and sequence numbers. Note that the authentication procedure can also be relaxed so as to accept all the attach procedures.

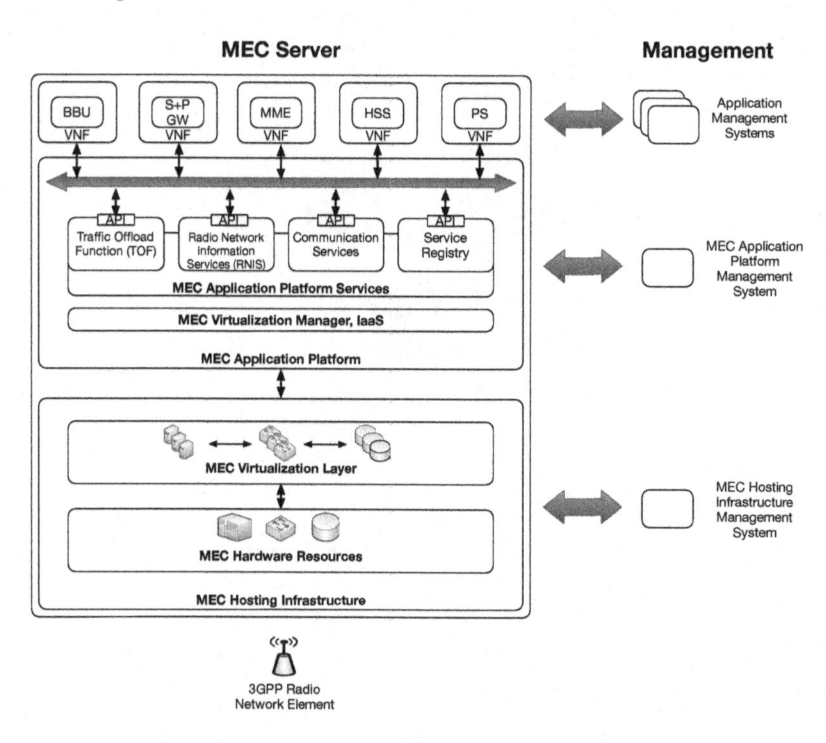

Figure 11.4. *The architecture of the DTN/ICN PS solution*

All the VNF functions are instantiated on the same edge cloud. The BBU has to be re-instantiated to acknowledge local copies of the MME, SPGW and HSS providing core network services. The MME, SPGW and HSS are minimal services of a small footprint. They provide basic LTE functions and connect UEs attached with the macro eNB. Due to the basic core function, the UEs attached to the same eNB can communicate directly with the help of the PS VNF. The PS VNF is based upon DTN and/or ICN applications such as CCNx[1] or DTN2[2]. It is a communication

1 http://www.blogs.parc.com/ccnx/ccnx-downloads/

2 https://www.sourceforge.net/projects/dtn/files/DTN2/

endpoint and a relay between other clients instantiated on UEs. The established setup allows end users to attach to macro eNB (see Figure 11.6). The rescue teams can now freely attach to the open BTS instantiated and exchange data using DTN and/or ICN relay points. If a macro eNB shares a functional X2 interface with another nearby base station, the X2 can be used as an interface to share data among nearby cells. Otherwise, the cellcell communication can be based on DTN with data mules.

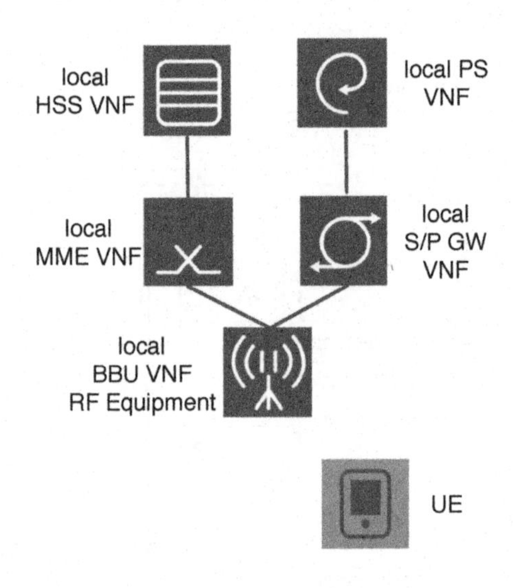

Figure 11.5. *Service bundle for PS applications*

DTN could enable a push communication between isolated eNBs when a disaster has occurred. In particular, in disaster scenarios with no end-to-end communication, DTN can provide access to isolated areas through its main characteristics, which is delay tolerance. To this end we can consider this as a MANET, where users are equipped with devices (smartphones, tablets) and could potentially be used as data mules between the isolated nodes and the rescue teams, i.e. as shown in Figure 11.6; in this case there is no functional interface between nearby cells, mobile users in an area could therefore store data in their UE (i.e. mobile phones, tablets) and act as data mules transferring information from one cell to another.

ICN could perform a similar functionality. Although users are used as data mules, content is exchanged based on information in a pull manner. This is advantageous, since an isolated cell could not know the exact location of a functional cell and could just transmit data based on the content, not on the location.

When connectivity to the regular core network restores, the service bundle is removed, and the local management entity establishes the original BBU again.

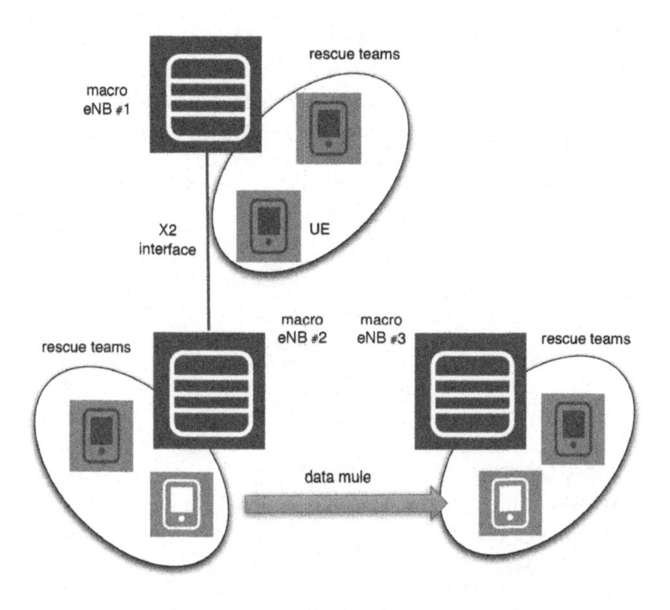

Figure 11.6. *Service bundle functionality*

11.4. Example implementation

11.4.1. *Juju VNF Manager*

The main building block of our system is Juju developed by Canonical. Juju is a domain neutral mechanism, which provides a generic VNF Manager (VNFM) that can be adapted to heterogeneous environments such as Infrastructure as a Service (IaaS) and Platform as a Service (PaaS) clouds. It natively supports service provisioning and scaling functions for scale-in/scale-out scenarios, therefore dynamically handling workloads by properly adjusting resources to momentary situations.

Juju provisions various services provided as software on-demand. It spans a large variety of applications such as databases (e.g. MySql), messaging systems (e.g. RabbitMQ) and monitoring infrastructures (e.g. Zabbix, Nagios) known from classical networks, but also more recently LTE network functions such as software-based BBUs, EPCs and HSS delivered by EURECOM as OpenAirInterface [NIK 14].

Services are described by charms that are service manifests allowing for appropriate service configurations. Juju allows for "gluing" or "bundling" services all together by implementing logic allowing for automatic associations between services (i.e. service chaining).

In the ETSI Management and Orchestration (MANO)[3] architecture, Juju should be classified as a VNFM of extended capabilities, helping MANO vendors to implement advanced business logic in the service orchestration part, to support an enhanced Quality of Service (QoS) through contracting appropriate Service Level Agreements (SLAs).

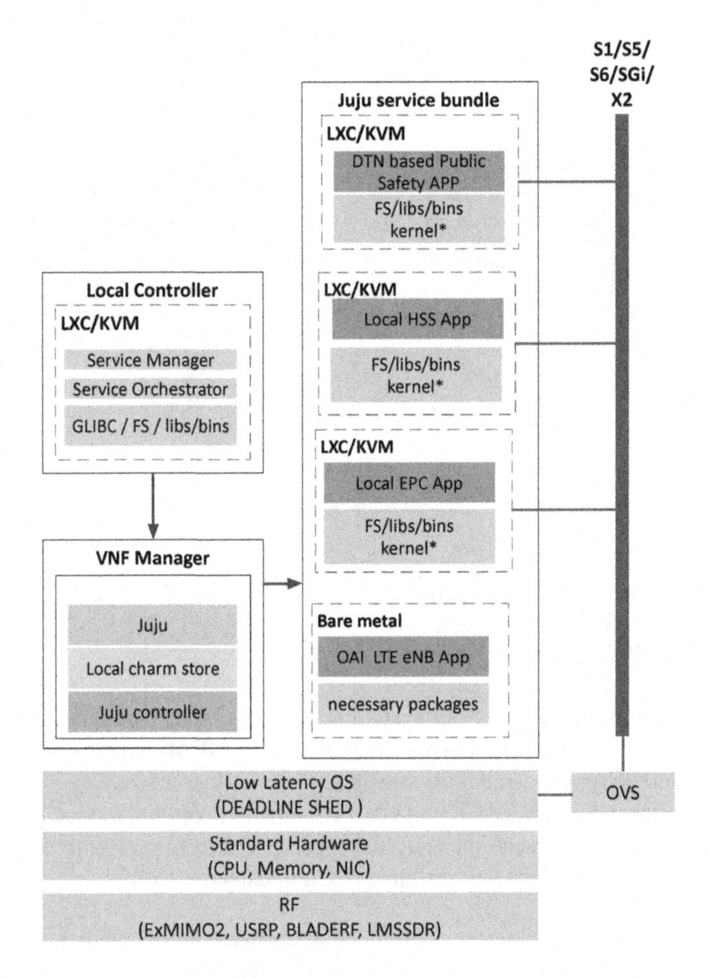

Figure 11.7. *Software architecture*

The hardware/software architecture of the system is presented in Figure 11.7. At the bottom, we illustrate hardware resources (i.e. RF hardware such as

3 http://www.osm.etsi.org

ExpressMIMO2, Ettus USRP; computing infrastructure) used by the MEC. The following layer is the operating system of the host controlling the hardware, and providing network and computing virtualization tools (OVS[4], KVM[5], XEN[6], LXC[7], etc.). Juju, together with local charm store and Juju controller, plays the role of the VNFM, which allows spawning VNF bundles on the MEC infrastructure. The local controller instantiated through Juju is a VNF that monitors the health of the network, and establishes the service bundle for PS applications when the eNB running with a disconnected core is detected through a telemetry service. The Juju service bundle for PS contains the whole LTE network stack, and a PS VNF connected with the help of the Open Virtual Switch (OVS) providing a virtual network.

11.4.2. *OVS virtual switch*

The OVS is a crucial element of the designed setup. The main concern of the virtual switch is put on delay and throughput of the virtual infrastructure. To confirm network delay in the virtual environment, we have created a laboratory setup using two computers directly connected with 1 GbE-T Ethernet cards in the laboratory of the University of Bern. They both use Intel i7 CPU (host1: 4 cores, 3.6 GHz, 32 GB RAM, Ubuntu 16.04, Kernel 4.4.0) and (host2: 2 cores, 3.4 GHz, 16 GB RAM, Ubuntu 16.04, Kernel 4.4.0) respectively. The faster machine, host1, is used to test the capacity of the virtual switch and various virtualization methods. We have tested the following configurations:

– direct communication between two physical hosts with physical interfaces (A);

– direct communication between two physical hosts; host1 has its physical interface configured as a port of an OVS switch. Two implementations of OVS are used: a regular one (B) and Intel dpdk (C);

– Linux Container (LXC) instantiated on host1, communicates with host2. The physical interface of host1 is configured as a port of an OVS switch. Two implementations of OVS are used, again a regular one (D) and dpdk (E);

– KVM instance on host1, communicates with host2. The physical interface of host1 is configured as a port of an OVS switch. Two implementations of OVS are used, again a regular one (F) and dpdk (G). In the regular mode, KVM uses TUN/TAP virtual interfaces configured as ports of the OVS. In dpdk mode, special dpdkvhostusers interfaces are used instead.

4 http://www.openvswitch.org/

5 http://www.linux-kvm.org

6 http://www.xenproject.org

7 http://www.linuxcontainers.org

In modes (except dpdk), the OVS is able to saturate the 1 GbE-T link between host1 and host2 providing throughput of about 89 MB/s ± 5 MB/s. The dpdk mode displayed a lower performance of about 60 MB/s ± 1 MB/s; however, more tests with different equipment are required to fully confirm this finding. The dpdk mode provides a dramatic improvement in the average communication delays (see Figure 11.8) especially in KVM. The round-trip delays were established by sending 64 bytes packets (ping) between communicating entities. The best performance of the virtual environment was established in mode (E), i.e. LXC with the dpdk OVS switch. It was then shortly followed by the KVM with dpdkvhostusers interfaces and dpdk OVS switch. We therefore see that the optimization of delay and throughput in OVS is getting closer to the performance of the physical environment.

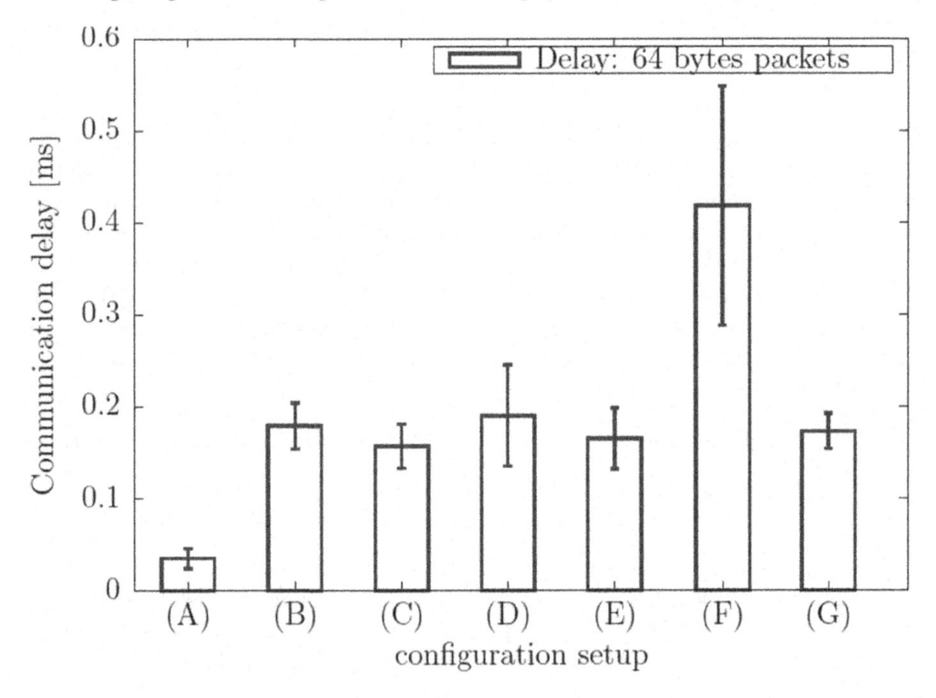

Figure 11.8. *Delay of the OVS switch*

11.4.3. *PS bundle at the network edge*

Due to the LTE network stack containing a BBU, EPC and HSS, the UE of the user can establish a connection with the network and connect to the PS application.

In Figure 11.9, we gather instantiation times for the PS service bundle (MySql is a supporting system for HSS). We tested two scenarios, where DTN (i.e. PS instance),

MySql[8], EPC[9] and HSS[10] were instantiated on KVM and LXC respectively. In both scenarios, however, the eNB[11] runs on the host as a bare metal service. This is to simplify the setup as in the bare metal mode, a pass-through between RF equipment and the container/VM is not necessary and the BBU can enjoy real-time capabilities of the host kernel. We use a single host machine with Intel 3.20 GHz quad core CPU and 16 GB RAM. The services use 1 thread, 1 GB RAM; 1 thread, 1 GB RAM; 4 threads, 8 GB RAM; 1 thread, 2 GB RAM; 1 thread, 1 GB RAM for MySql, HSS, eNB, EPC and DTN respectively.

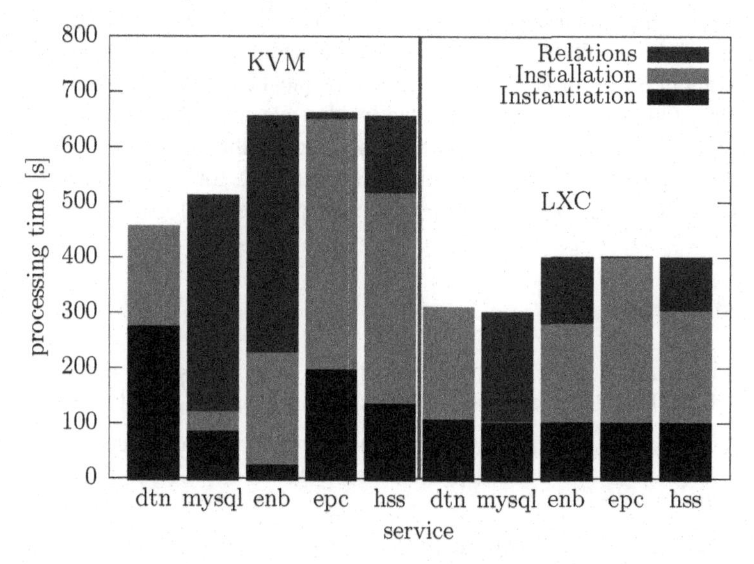

Figure 11.9. *Instantiation time*

Service provisioning contains three phases: instantiation (bare metal, LXC container, KVM VM), software installation and relation establishment. In the relation establishment, the service installed earlier, mostly waits for its counterpart services to become ready. A reconfiguration is performed when another associated service becomes operational. For example, MySql waits for HSS to become ready. When this happens, both services bundle together and instantiation of MySql finishes. The eNB, EPC, HSS and DTN are not provided with binary packages, but compiled upon instantiation, therefore long installation times are expected. However, the complete PS stack requires about 650 s on KVM and 400 s on LXC to become

8 https://www.jujucharms.com/mysql/55

9 https://www.jujucharms.com/u/navid-nikaein/oai-epc/trusty/27

10 https://www.jujucharms.com/u/navid-nikaein/oai-hss/trusty/16

11 https://www.jujucharms.com/u/navid-nikaein/oai-enb/trusty/20

fully provisioned. A UE can connect to the service (e.g. DTN in our scenario) and request data. We also noticed that on LXC, Juju creates containers in parallel, but deployment of VMs on KVM seems to be sequential.

11.5. Conclusion

We have developed a PS architecture for a macro eNB with MEC operating in a disconnected core scenario. MEC successfully establishes the whole LTE stack and provisions necessary PS services such as a DTN agent at the edge. The instantiation time is of about 650 s for KVM and 400 s on a typical commodity hardware, which is reasonable. Our solution can significantly improve the PS communication at the network edge. Future works include packaging of the modules, thus improving instantiation time. Moreover, authentication procedures for the PS scenarios in LTE have to be worked out. We will also target other radio technologies such as WiFi.

11.6. Bibliography

[BUR 06] BURGESS J., GALLAGHER B., JENSEN D. *et al.*, "MaxProp: routing for vehicle-based disruption-tolerant networks", *INFOCOM*, vol. 6, pp. 1–11, 2006.

[DEM 07] DEMMER M., FALL K., "DTLSR: delay tolerant routing for developing regions", *Proceedings of the 2007 Workshop on Networked Systems for Developing Regions*, ACM, p. 5, 2007.

[FAJ 14] FAJARDO J.T.B., YASUMOTO K., SHIBATA N. *et al.*, "Disaster information collection with opportunistic communication and message aggregation", *Journal of Information Processing*, vol. 22, no. 2, pp. 106–117, 2014.

[FAV 16] FAVRAUD R., APOSTOLARAS A., NIKAEIN N. *et al.*, "Toward moving public safety networks", *IEEE Communications Magazine*, vol. 54, no. 3, pp. 14–20, 2016.

[LIN 03] LINDGREN A., DORIA A., SCHELÉN O., "Probabilistic routing in intermittently connected networks", *ACM SIGMOBILE Mobile Computing and Communications Review*, vol. 7, no. 3, pp. 19–20, 2003.

[NIK 14] NIKAEIN N., KNOPP R., KALTENBERGER F. *et al.*, "Demo: OpenAirInterface: an open LTE network in a PC", *Proceedings of the 20th Annual International Conference on Mobile Computing and Networking*, MobiCom'14, ACM, New York, pp. 305–308, 2014.

[PAT 14] PATEL M., NAUGHTON B., CHAN C. *et al.*, "Mobile-Edge Computing – Introductory Technical White Paper", ETSI Technical White Paper, 2014.

[SOU 16] SOUSA B., CORDEIRO L., SIMOES P. *et al.*, "Toward a fully cloudified mobile network infrastructure", *IEEE Transactions on Network and Service Management*, vol. 13, no. 3, pp. 547–563, 2016.

[SPY 05] SPYROPOULOS T., PSOUNIS K., RAGHAVENDRA C.S., "Spray and wait: an efficient routing scheme for intermittently connected mobile networks", *Proceedings of the 2005 ACM SIGCOMM Workshop on Delay-Tolerant Networking*, ACM, pp. 252–259, 2005.

[TRO 16] TRONO E.M., FUJIMOTO M., SUWA H. *et al.*, "Disaster area mapping using spatially-distributed computing nodes across a DTN", *IEEE International Conference on Pervasive Computing and Communication Workshops (PerCom Workshops)*, pp. 1–6, 2016.

[TYS 14] TYSON G., BODANESE E., BIGHAM J. *et al.*, "Beyond content delivery: Can ICNs help emergency scenarios?", *IEEE Network*, vol. 28, no. 3, pp. 44–49, 2014.

[UCH 13] UCHIDA N., KAWAMURA N., WILLIAMS N. *et al.*, "Proposal of delay tolerant network with cognitive wireless network for disaster information network system", *27th International Conference on Advanced Information Networking and Applications Workshops (WAINA)*, pp. 249–254, 2013.

[VAH 00] VAHDAT A., BECKER D. *et al.*, "Epidemic routing for partially connected ad-hoc networks", Technical Report CS-200006, Duke University, 2000.

List of Authors

Akram AL-HOURANI
School of Electrical and Computer Engineering
RMIT University
Melbourne
Australia

Ludovic APVRILLE
System-on-Chip Laboratory (LabSoC)
Institut Telecom ParisTech
Sophia Antipolis
France

Torsten BRAUN
Institute of Computer Science
University of Bern
Switzerland

Xavier CALDERÓN
Department of Telecommunications and Information Networks
Escuela Politécnica Nacional
Quito
Ecuador

Daniel CÂMARA
Central Service of Criminal Intelligence
French National Gendarmerie
Pontoise
France

Sandip CHAKRABORTY
Indian Institute of Technology Kharagpur
Kharagpur, West Bengal
India

Sathyanarayanan CHANDRASEKHARAN
School of Electrical and Computer Engineering
RMIT University
Melbourne
Australia

Sajal DAS
Missouri University of Science and Technology
Rolla, Missouri
USA

Tomaso DE COLA
Institute of Communications and Navigation
DLR German Aerospace Center
Cologne
Germany

Almudena DÍAZ-ZAYAS
University of Málaga
Andalucia Tech
Málaga
Spain

Georgios ELLINAS
KIOS Research
Center for Intelligent Systems and Networks and the Department of Electrical and Computer Engineering
University of Cyprus
Nicosia
Cyprus

Manuel ESTEVE
Department of Communications
Universidad Politécnica de Valencia
Valencia
Spain

César Agusto GARCÍA-PÉREZ
University of Málaga
Andalucia Tech
Málaga
Spain

André GOMES
Institute of Computer Science
University of Bern
Switzerland

Krishnandu HAZRA
National Institute of Technology
Durgapur
Durgapur, West Bengal
India

Abbas JAMALIPOUR
School of Electrical and Information Engineering
University of Sydney
Australia

Sebastian JARAMILLO
Department of Telecommunications and Information Networks
Escuela Politécnica Nacional
Quito
Ecuador

Eirini KALOGEITON
Institute of Computer Science
University of Bern
Switzerland

Sithamparanathan KANDEEPAN
School of Electrical and Computer Engineering
RMIT University
Melbourne
Australia

Kostas KATSALIS
Eurecom
Sophia Antipolis
France

Panayiotis KOLIOS
KIOS Research
Center for Intelligent Systems and Networks and the Department of Electrical and Computer Engineering
University of Cyprus
Nicosia
Cyprus

Shivsubramani KRISHNAMOORTHY
Amrita University
Tamilnadu
India

Letitia W. LI
System-on-Chip Laboratory
(LabSoC)
Institut Telecom ParisTech
Sophia Antipolis
France

Pedro MERINO
University of Málaga
Andalucia Tech
Málaga
Spain

Donal MORRIS
Redzinc Services Limited
Dublin
Ireland

Javier MULERO CHAVES
Institute of Communications and
Navigation
DLR German Aerospace Center
Cologne
Germany

Subrata NANDI
National Institute of Technology
Durgapur
Durgapur, West Bengal
India

Navid NIKAEIN
EURECOM
Sophia Antipolis
France

Carlos PALAU
Department of Communications
Universidad Politécnica de Valencia
Valencia
Spain

Christos PANAYIOTOU
KIOS Research
Center for Intelligent Systems and
Networks and the Department of
Electrical and Computer Engineering
University of Cyprus
Nicosia
Cyprus

Partha Sarathi PAUL
National Institute of Technology
Durgapur
Durgapur, West Bengal
India

Prabaharan POORNACHANDRAN
Amrita University
Tamilnadu
India

Álvaro RÍOS
University of Málaga
Andalucia Tech
Málaga
Spain

Sujay SAHA
National Institute of Technology
Durgapur
Durgapur, West Bengal
India

Eryk SCHILLER
Institute of Computer Science
University of Bern
Switzerland

Sujadevi VIJAYA GANGADHARAN
Amrita University
Tamilnadu
India

Ana Maria ZAMBRANO
Department of Telecommunications and Information Networks
Escuela Politécnica Nacional
Quito
Ecuador

Oscar Marcelo ZAMBRANO
Department of Communications
Universidad Politécnica de Valencia
Valencia
Spain

Index

Printed in the United States
By Bookmasters